Environmental Hydrology

Environmental Hydrology

Shaun Grantham

Larsen & Keller
www.larsen-keller.com

Environmental Hydrology
Shaun Grantham
ISBN: 978-1-64172-078-6 (Hardback)

⊟ Larsen & Keller

Published by Larsen and Keller Education,
5 Penn Plaza,
19th Floor,
New York, NY 10001, USA

Cataloging-in-Publication Data

Environmental hydrology / Shaun Grantham.
       p. cm.
Includes bibliographical references and index.
ISBN 978-1-64172-078-6
1. Ecohydrology. 2. Hydrology. 3. Hydraulic engineering--Environmental aspects. I. Grantham, Shaun.
QH541.15.E19 E58 2019
551.48--dc23

For more information regarding Larsen and Keller Education and its products, please visit the publisher's website www.larsen-keller.com

Table of Contents

Permissions

Index

Preface

Hydrology is the science concerned with the study of the movement and distribution of water on the Earth. The study of water resources, the water cycle, watershed sustainability and analysis of the quality of water are also within this domain. This field subdivides into the domains of groundwater, surface water and marine hydrology. These domains branch into a number of significant sub-disciplines, such as chemical hydrology, hydrogeology, isotope hydrology, ecohydrology, etc. The applications of hydrological studies are in the calculation of rainfall and surface runoff, determination of the water balance and agricultural water balance of a region, prediction and mitigation of floods and landslides, provision of potable water, etc. This book is a compilation of chapters that discuss the most vital concepts in the field of environmental hydrology. Some of the diverse topics covered herein address the varied branches that fall under this category. Coherent flow of topics, student-friendly language and extensive use of examples make this book an invaluable source of knowledge.

A detailed account of the significant topics covered in this book is provided below:

Chapter 1, The distribution, movement and quality of water on Earth is studied under hydrology. It also involves the study of water resources, the water cycle and environmental watershed sustainability. The aim of this chapter is to provide an introduction to hydrology through the elucidation of topics related to the distribution of water on Earth and bodies of water. **Chapter 2**, The hydrological cycle, also known as the water cycle, refers to the continuous movement of water between the reservoirs of fresh and saline water, ice and atmospheric water. This cycle is aided by the processes of evaporation, condensation, precipitation, subsurface flow, etc. This chapter closely examines the crucial aspects of hydrological cycle and hydrological flow, such as streamflow, environmental flow, freshwater inflow, water balance, etc. **Chapter 3**, The transformation of water from a liquid phase to a gaseous phase during its movement from water bodies to the atmosphere is known as evaporation. The process of the release of water vapor from the soil and plants into the atmosphere is called transpiration. A detailed analysis of the fundamental concepts of evaporation and transpiration has been provided in this chapter, which includes topics such as pan evaporation, potential evaporation, transpiration and evapotranspiration. **Chapter 4**, The flow of excess meltwater and stormwater over the surface of the Earth is known as surface runoff. It is a significant component of the hydrological cycle. This chapter discusses diverse aspects of surface runoff and the various processes of snowmelt, rainfall runoff, stormwater runoff, urban runoff, etc. **Chapter 5**, The process by which any surface water seeps into the soil is known as infiltration hydrology. It is caused due to the gravity and capillary action. Infiltration rate and capacity is affected by various factors such as vegetation cover, soil texture, soil temperature, rainfall intensity, etc. This is an important chapter, which analyzes infiltration hydrology, field capacity, soil plant atmosphere continuum, etc. **Chapter 6**, Hydrogeology is an area of geology which deals with the movement and distribution of groundwater in the soil and rocks of the Earth's surface. All the important aspects of hydrogeology such as groundwater flow and vadose zone have been extensively covered in this chapter. **Chapter 7**, Rivers and streams provide habitats for aquatic plants and animals. They also carry sediment and water from high elevated areas to downstream estuaries, lakes and oceans. The various aspects integral to natural stream processes include stream flow, stream stability, stream restoration, etc. These have been extensively discussed in this chapter.

It gives me an immense pleasure to thank our entire team for their efforts. Finally in the end, I would like to thank my family and colleagues who have been a great source of inspiration and support.

Shaun Grantham

Hydrology and Water Resources

The distribution, movement and quality of water on Earth is studied under hydrology. It also involves the study of water resources, the water cycle and environmental watershed sustainability. The aim of this chapter is to provide an introduction to hydrology through the elucidation of topics related to the distribution of water on Earth and bodies of water.

Hydrology is the study of the movement, distribution, and quality of water throughout the Earth. It addresses both the hydrologic cycle and water resources. A practitioner of hydrology, or *hydrologist*, may work in any of several fields: earth science, environmental science, physical geography, civil engineering, and environmental engineering.

Hydrological research is useful in that it allows engineers to (a) design irrigation schemes, water-supply systems, dams, bridges, and sewers; (b) predict and mitigate the risk of floods, droughts, landslides, erosion, and sedimentation; and (c) assess the risk of contaminant transport. In this manner, it provides insights for environmental engineering, policy, and planning.

Hydrology has been a subject of investigation and engineering for millennia. For example, around 4000 B.C.E., the Nile was dammed to improve agricultural productivity of previously barren lands. Mesopotamian towns were protected from flooding with high earthen walls. Aqueducts were built by the Greeks and Romans, while the Chinese built irrigation and flood control works.

In the first century B.C.E., Marcus Vitruvius described a philosophical theory of the hydrologic cycle, according to which precipitation falling in the mountains infiltrated the Earth's surface and led to streams and springs in the lowlands. With the adoption of a more scientific approach, Leonardo da Vinci and Bernard Palissy independently reached an accurate representation of the hydrologic cycle. It was not until the seventeenth century that hydrologic variables began to be quantified.

Pioneers of the modern science of hydrology include Pierre Perrault, Edme Mariotte, and Edmund Halley. By measuring rainfall, runoff, and drainage area, Perrault showed that rainfall was sufficient to account for flow of the Seine. Marriotte combined velocity and river cross-section measurements to obtain discharge, again in the Seine. Halley showed that evaporation from the Mediterranean Sea was sufficient to account for the outflow of rivers flowing into the sea.

Advances in the eighteenth century included the Bernoulli equation and piezometer by Daniel Bernoulli, the Pitot tube, and the Chezy formula. The nineteenth century saw development in groundwater hydrology, including Darcy's law, the Dupuit-Thiem well formula, and Hagen-Poiseuille's capillary flow equation.

Rational analyses began to replace empiricism in the twentieth century, while governmental agencies began their own hydrological research programs. Of particular importance were Leroy Sherman's unit hydrograph, the infiltration theory of Robert E. Horton, and C.V. Theis's equation describing well hydraulics.

Since the 1950s, hydrology has been approached with a more theoretical basis than in the past, facilitated by advances in the physical understanding of hydrological processes and the advent of computers.

Hydrologic Cycle and Transport

The central theme of hydrology is that water moves throughout the Earth by different pathways and at different rates. The most striking image of this is in the evaporation of water from the ocean, to form clouds. These clouds drift over the land and produce rain. The rainwater flows into lakes, rivers, or aquifers. The water in lakes, rivers, and aquifers then either evaporates into the atmosphere or eventually flows back to the ocean, completing a cycle.

Moreover, water movement is a significant means by which other material, such as soil or pollutants, is transported from place to place. The initial input in receiving waters may arise from a point source discharge or a line source or area source, such as surface runoff. Since the 1960s, rather complex mathematical models have been developed, facilitated by the availability of high speed computers. The most common pollutant classes analyzed are nutrients, pesticides, total dissolved solids, and sediment.

Branches of Hydrology

- *Chemical hydrology* is the study of the chemical characteristics of water. It examines how water is affected as it comes into contact with different materials on and below the Earth›s surface. This field includes studies on the mechanisms by which salts are transported by such processes as erosion, runoff, evaporation, and precipitation.

- *Ecohydrology* is the study of ecological processes in the hydrologic cycle. As these processes occur in the soil and plant foliage, ecohydrologists study how the hydrologic system affects plant physiology, soil moisture, and plant diversity and spatial orientation in various regions over a period of time. Ecohydrology has four main components: infiltration of precipitation into the soil, evapotranspiration, leakage of water into deeper portions of the soil not accessible to the plant, and runoff from the ground surface.

- *Hydrogeology* (or *geohydrology*) is the study of the distribution and movement of water in aquifers and shallow porous media that is, the porous layers of rock, sand, silt, and gravel below the Earth's surface. Hydrogeology examines the rate of diffusion of water through these media as the water moves down its energy gradient. The flow of water in the shallow subsurface is also pertinent to the fields of soil science, agriculture, and civil engineering. The flow of water and other fluids (hydrocarbons and geothermal fluids) in deeper formations is relevant to the fields of geology, geophysics, and petroleum geology.

- *Hydroinformatics* is the adaptation of information technology to hydrology and water resources applications. Its purpose is to facilitate decision-making for many critical applications. Hydrological data are collected, stored, processed, and analyzed using modeling techniques and simulations, based on the knowledge of particular systems. Three common types of hydrological data collected are: the rate of flow of major rivers and streams, precipitation, and water height in wells.

- *Hydrometeorology* is the study of the transfer of water and energy between land and water body surfaces and the lower atmosphere. Hydrometeorology incorporates meteorology to solve hydrological problems. These problems include forcasting flood or drought, or determining water resources and the safety of dams. Hydrometeorologists try to determine, through empirical data or theory, how the dynamics of water in the atmosphere affect the greatest levels of precipitation reaching the ground. The domain of hydrometeorology in the physical sciences is not very clearly defined, as it involves cloud physics, climatology, weather forecasting, and hydrology, to name a few.

- *Hydromorphology* is the study of the physical characteristics of bodies of water on the Earth›s surface, including river basins, channels, streams, and lakes. Water quality, levels of pollution, and biological components needed for ecological system maintenance are a few areas assessed when classifying water systems. Hydromorphology studies the dynamics of groundwater flow into channels, lakes, and streams. It measures flow patterns and geometry as well as routing flows to avoid flooding or drought.

- *Isotope hydrology* is the study of the isotopic signatures of water. This subfield of hydrology utilizes isotopic dating to determine the origin and age of water throughout its movement within the hydrologic cycle. Isotopic dating involves measuring the levels of deviation in the isotopes of oxygen and hydrogen in water. Researchers are able to determine groundwater dated as far back as the Ice Age by using these techniques. Isotope hydrology deals with water usage policy, mapping aquifers, conservation of water resources, and maintaining pollution levels. One way isotopic hydrology is applied today is in the mitigation of arsenic levels in the drinking water of Bangladesh.

- *Surface-water hydrology* is the study of bodies of water on or near the Earth's surface. Rivers, dams, lakes, and reservoirs are all part of this area of study, which further includes the systems used in recreational activity and transportation. Surface hydrology addresses issues pertaining to eroding soils and streams due to surface flow. Flooding, nutrient runoff, and pollutants are a few of the effects addressed, as well as the destruction of civil constructions such as dams. Methods of hydraulic and hydrologic design regulation are also undertaken in this field of study, as researchers simulate the long and short-term effects of anthropogenically manipulated surface water forms.

Related Fields

- Aquatic chemistry: Aquatic chemistry studies chemical reactions in aqueous solutions, including acid-base reactions, oxidation-reduction reactions, precipitation reactions, and dissolution reactions. It can be applied to addressing issues on water pollution and treatment and creating sustainable methods of production with little environmental impact.

- Civil engineering: Related to the study of hydrology, civil engineers contribute to the planning, design, construction, and maintenance of structures associated with hydraulics. For instance, the engineers are involved in controlling water flow by way of draining swamps, municipal sewage disposal, flood control, and irrigation. They also work in creating structures that help route water flow in dams and bridges.

- Climatology: Climatology is the study of climate, which is scientifically defined as weather conditions averaged over a period of time. It is a branch of the atmospheric sciences. Average precipitation and temperature trends are measured over specific regions.

- Environmental engineering: Environmental engineering combines science and engineering principles to address ways in which to improve the quality of air, land, and water for living organisms. Chemical, biological, and geological sciences are incorporated into the techniques of mechanical, civil, and chemical engineering to address issues of public health and policy. Remediation of polluted sites, sanitary engineering, and waste reduction and prevention are keys areas of concern.

- Physical Geography: Physical geography deals with topics concerning the Earth's surface, including glaciers, landforms, rivers and oceans, climate, and hydrological processes driven by the Sun. It involves the systematic study of patterns in the biosphere, lithosphere, hydrosphere, and atmosphere.

- Geomorphology: Geomorphology is the study of landforms, including their origin and evolution, and the processes that shape them. A combination of field observation, physical experiment, and numerical modeling help geomorphologists to understand landform history and dynamics, and predict future changes. Applications of geomorphology include landslide prediction and mitigation, river control and restoration, coastal protection, and assessing the presence of water on Mars.

- Hydraulic engineering: Hydraulic engineering is a subdiscipline of civil engineering that focuses on the flow and conveyance of fluids, particularly water. It involves the design and construction of hydraulic structures such as bridges, dams, canals, channels, and levees, and also aligns itself with the goals of sanitary and environmental engineering.

- Limnology: Limnology involves the study of inland waters, both saline and fresh. Specifically, it is the study of lakes, ponds, and rivers (natural and manmade), including their biological, physical, chemical, and hydrological aspects.

- Oceanography: Oceanography is the study of the Earth's seas and oceans. It includes the geological movement of tectonic platesunder the Earth's surface, physical oceanographic characteristics, chemical processes, and marine biological processes taking place in these bodies of water.

Hydrologic Measurements

The movement of water through the Earth can be measured in a number of ways. This information is important for both assessing water resources and understanding the processes involved in the hydrologic cycle. The following is a list of devices used by hydrologists and what they are used to measure.

- Disdrometer - precipitation characteristics

- Symon's evaporation pan - evaporation

- Infiltrometer - infiltration

- Piezometer - groundwater pressure and, by inference, groundwater depth

- Radar - cloud properties

- Rain gauge - rain and snowfall

- Satellite - topographic patterns of surface water

- Sling psychrometer - humidity

- Stream gauge - stream flow

- Tensiometer - soil moisture

- Time domain reflectometer - soil moisture

Hydrologic Prediction

Observations of hydrologic processes are used to make predictions of future water movement and quantity.

Statistical Hydrology

By analyzing the statistical properties of hydrologic records, such as rainfall or river flow, hydrologists can estimate future hydrologic phenomena. This approach, however, assumes that the characteristics of the processes remain unchanged.

These estimates are important for engineers and economists so that they can perform proper risk analysis for future decisions in infrastructure and to determine the yield reliability characteristics of water supply systems. Statistical information is utilized to formulate operating rules for large dams that are part of systems set up to meet agricultural, industrial, and residential demands.

Hydrologic Modeling

Hydrologic models are simplified, conceptual representations of a part of the hydrologic cycle. They are primarily used for hydrologic prediction and for understanding hydrologic processes. Two major types of hydrologic models can be distinguished:

- Models based on data: These models use mathematical and statistical concepts to link a certain input (such as rainfall) to the model output (such as runoff). These models are known as stochastic hydrology models.

- Models based on process descriptions: These models try to represent the physical processes observed in the real world. Typically, they contain representations of surface runoff, subsurface flow, evapotranspiration, and channel flow, but they can be far more complicated. These models are known as deterministic hydrology models.

Applications of Hydrology

- Mitigating and predicting the risk of flood, landslide, and drought.

- Designing irrigation schemes and managing agricultural productivity.

- Providing drinking water.

- Designing dams for water supply or hydroelectric power generation.

- Designing bridges.

- Designing sewers and urban drainage system.

- Analyzing the impacts of antecedent moisture on sanitary sewer systems.

- Predicting geomorphological changes, such as erosion or sedimentation.

- Assessing the impacts of natural and anthropogenic environmental change on water resources.

- Assessing contaminant transport risk and establishing environmental policy guidelines.

Water

Water is a substance composed of the chemical elements hydrogen and oxygen and existing in gaseous, liquid, and solid states. It is one of the most plentiful and essential of compounds. A tasteless and odourless liquid at room temperature, it has the important ability to dissolve many other substances. Indeed, the versatility of water as a solvent is essential to living organisms. Life is believed to have originated in the aqueous solutions of the world's oceans, and living organisms depend on aqueous solutions, such as blood and digestive juices, for biological processes. In small quantities water appears colourless, but water actually has an intrinsic blue colour caused by slight absorption of light at red wavelengths.

Although the molecules of water are simple in structure (H_2O), the physical and chemical properties of the compound are extraordinarily complicated, and they are not typical of most substances found on Earth. For example, although the sight of ice cubes floating in a glass of ice water is commonplace, such behaviour is unusual for chemical entities. For almost every other compound, the solid state is denser than the liquid state; thus, the solid would sink to the bottom of the liquid. The fact that ice floats on water is exceedingly important in the natural world, because the ice that forms on ponds and lakes in cold areas of the world acts as an insulating barrier that protects the aquatic life below. If ice were denser than liquid water, ice forming on a pond would sink, thereby exposing more water to the cold temperature. Thus, the pond would eventually freeze throughout, killing all the life-forms present.

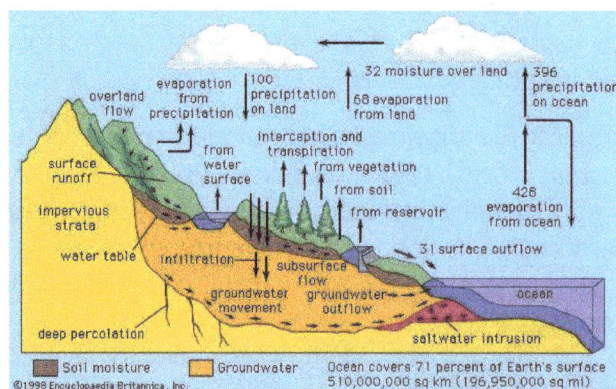

In the hydrologic cycle, water is transferred between the land surface, the ocean, and the atmosphere. The numbers on the arrows indicate relative water fluxes

Water occurs as a liquid on the surface of Earth under normal conditions, which makes it invaluable for transportation, for recreation, and as a habitat for a myriad of plants and animals. The fact that water is readily changed to a vapour (gas) allows it to be transported through the atmosphere from the oceans to inland areas where it condenses and, as rain, nourishes plant and animal life.

Because of its prominence, water has long played an important religious and philosophical role in human history. In the 6th century BCE, Thales of Miletus, sometimes credited for initiating Greek philosophy, regarded water as the sole fundamental building block of matter:

It is water that, in taking different forms, constitutes the earth, atmosphere, sky, mountains, gods and men, beasts and birds, grass and trees, and animals down to worms, flies and ants. All these are different forms of water. Meditate on water!

Two hundred years later, Aristotle considered water to be one of four fundamental elements, in addition to earth, air, and fire. The belief that water was a fundamental substance persisted for more than 2,000 years until experiments in the second half of the 18th century showed that water is a compound made up of the elements hydrogen and oxygen.

The water on the surface of Earth is found mainly in its oceans (97.25 percent) and polar ice caps and glaciers (2.05 percent), with the balance in freshwater lakes, rivers, and groundwater. As Earth's population grows and the demand for fresh water increases, water purification and recycling become increasingly important. Interestingly, the purity requirements of water for industrial use often exceed those for human consumption. For example, the water used in high-pressure boilers must be at least 99.999998 percent pure. Because seawater contains large quantities of dissolved salts, it must be desalinated for most uses, including human consumption.

The Hoover Dam on the Colorado River at the border of Nevada and Arizona demonstrates how natural resources of water can be harnessed for a variety of purposes, including human consumption, irrigation and industry

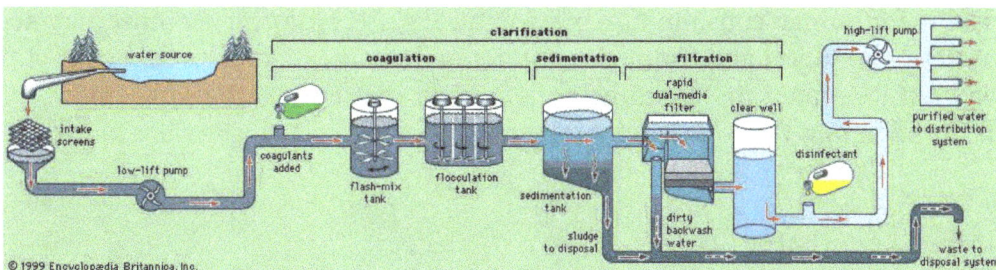

Water treatment systems are important for desalinating seawater so it can be used for human consumption and for purifying water for industrial use.

Surface Water

Most surface water comes from rainfall (precipitation) runoff from the surrounding land area (catchment). Of course not all runoff ends up in rivers, some evaporates, some is used by vegetation and part of it soaks into the ground recharging our groundwater systems, some of which can then seep back into the riverbeds. At a certain depth below the land surface, called the water table the ground becomes saturated with water. If a river happens to cut into this saturated layer, then water will seep out of the ground into the river. Groundwater seepage is most commonly seen in the form of springs eg. Berry Springs, Katherine Hot Springs and Bitter Springs.

Surface water includes larges rivers, ponds and lakes, and the small upland streams which may originate from springs and collect the run-off from the watersheds. The quantity of run-off depends upon a large number of factors, the most important of which are the amount and intensity of rainfall, the climate and vegetation and, also, the geological, geographical, and topographical features of the area under consideration. It varies widely, from about 20 % in arid and sandy areas where the rainfall is scarce to more than 50% in rocky regions in which the annual rainfall is heavy. Of the remaining portion of the rainfall. Some of the water percolates into the ground and the rest is lost by evaporation, transpiration and absorption.

The quality of surface water is governed by its content of living organisms and by the amounts of mineral and organic matter which it may have picked up in the course of its formation. As rain falls through the atmosphere, it collects dust and absorbs oxygen and carbon dioxide from the air. While flowing over the ground, surface water collects silt and particles of organic matter, some of which will ultimately go into solution. It also picks up more carbon dioxide from the vegetation and micro-organisms and bacteria from the topsoil and from decaying matter. On inhabited watersheds, pollution may include faecal material and pathogenic organisms, as well as other human and industrial wastes which have not been properly disposed of. In rural areas, water from small streams draining isolated or uninhabited watersheds may possess adequate bacteriological and chemical quality for human consumption in its natural state. However, in most instances surface water is subject to pollution and contamination by pathogenic organisms and cannot be considered safe without treatment. It should be remembered that clear water is not necessarily fit for human consumption and that one cannot depend wholly on self-purification to produce potable water.

There are three types of surface water:

Permanent (perennial) - permanent surface waters are present throughout the year. They are usually in the form of rivers, lakes, springs and swamps. At times when there is little or no rain, the water level is maintained by groundwater contributions.

Semi-permanent (ephemeral) - Semi-permanent water bodies are those that only hold water for part of the year. These are usually small creeks, lagoons, waterholes, or low lying areas in the arid zone.

Man-made – surface water can also be held in manmade structures ranging from lakes, dams and turkey nests to artificial swamps and sewage treatment ponds.

Underwater

Underwater refers to the region below the surface of water where the water exists in a swimming pool or a natural feature (called a body of water) such as an ocean, sea, lake, pond, or river.

Extent

Three quarters of the planet Earth is covered by water. A majority of the planet›s solid surface is abyssal plain, at depths between 4,000 and 5,500 metres (13,100 and 18,000 ft) below the surface of the oceans. The solid surface location on the planet closest to the centre of the orb is the Challenger Deep, located in the Mariana Trench at a depth of 10,924 metres (35,840 ft). Although a number of human activities are conducted underwater—such as research, scuba diving for work or recreation, or even underwater warfare with submarines, this very extensive environment on planet Earth is hostile to humans in many ways and therefore little explored. But it can be explored by sonar, or more directly via manned or autonomous submersibles. The ocean floors have been surveyed via sonar to at least a coarse resolution; particularly-strategic areas have been mapped in detail, in the name of detecting enemy submarines, or aiding friendly ones, though the resulting maps may still be classified.

Constraints on Non-aquatic Life

An immediate obstacle to human activity under water is the fact that human lungs cannot naturally function in this environment. Unlike the gills of fish, human lungs are adapted to the exchange of gases at atmospheric pressure, not liquids. Aside from simply having insufficient musculature to rapidly move water in and out of the lungs, a more significant problem for all air-breathing animals, such as mammals and birds, is that water contains so little dissolved oxygen compared with atmospheric air. Air is around 21% O_2; water typically is less than 0.001% dissolved oxygen.

The density of water also causes problems that increase dramatically with depth. The atmospheric pressure at the surface is 14.7 pounds per square inch or around 100 kPa. A comparable water pressure occurs at a depth of only 10 m (33 ft) (9.8 m (32 ft) for sea water). Thus, at about 10 m below the surface, the water exerts twice the pressure (2 atmospheres or 200 kPa) on the body as air at surface level.

For solids and liquids like bone, muscle and blood, this added pressure is not much of a problem; but it is a problem for any air-filled spaces like the mouth, ears, paranasal sinusesand lungs. This is because the air in those spaces reduces in volume when under pressure and so does not provide those spaces with support against the higher outside pressure. Even at a depth of 8 ft (2.4 m) underwater, an inability to equalize air pressure in the middle ear with outside water pressure can cause pain, and the tympanic membrane (eardrum) can rupture at depths under 10 ft (3 m). The danger of pressure damage is greatest in shallow water because the ratio of pressure change is greatest near the surface of the water. For example, the pressure increase between the surface and 10 m (33 ft) is 100% (100 kPa to 200 kPa), but the pressure increase from 30 m (100 ft) to 40 m (130 ft) is only 25% (400 kPa to 500 kPa).

Buoyancy

Any object immersed in water is provided with a buoyant force that counters the force of gravity, appearing to make the object less heavy. If the overall density of the object exceeds the density of water, the object sinks. If the overall density is less than the density of water, the object rises until it floats on the surface.

Note the bluish cast given to objects in this underwater photo of pillow lava (NOAA).

Photic Zone

With increasing depth underwater, sunlight is absorbed, and the amount of visible light diminishes. Because absorption is greater for long wavelengths (red end of the visible spectrum) than for short wavelengths (blue end of the visible spectrum), the colour spectrum is rapidly altered with increasing depth. White objects at the surface appear bluish underwater, and red objects appear dark, even black. Although light penetration will be less if water is turbid, in the very clear water of the open ocean less than 25% of the surface light reaches a depth of 10 m (33 feet). At 100 m (330 ft) the light present from the sun is normally about 0.5% of that at the surface.

The euphotic depth is the depth at which light intensity falls to 1% of the value at the surface. This depth is dependent upon water clarity, being only a few metres underwater in a turbid estuary, but may reach up to 200 metres in the open ocean. At the euphotic depth, plants (such as phytoplankton) have no net energy gain from photosynthesis and thus cannot grow.

At depths greater than a few hundred metres, the sun has little effect on water temperature, because the sun's energy has been absorbed by water at the surface. In the great depths of the ocean the water temperature is very low. In fact, 75% of the water in the world ocean (the great depths) has a temperature between 0 °C and 2 °C.

Conductivity

Water conducts heat around 25 times more efficiently than air. Hypothermia, a potentially fatal condition, occurs when the human body's core temperature falls below 35 °C. Insulating the body's warmth from water is the main purpose of diving suits and exposure suits when used in water temperatures below 25 °C.

Sound is transmitted about 4.3 times faster in water (1,484 m/s in fresh water) as it is in air (343 m/s). The human brain can determine the direction of sound in air by detecting small differences in the time it takes for sound waves in air to reach each of the two ears. For these reasons divers find it difficult to determine the direction of sound underwater. However, some animals have adapted to this difference and many use sound to navigate underwater.

Aquifer

An aquifer is a large underground storage space for water. They can be located right at the ground surface or very deep underground where they are impossible to access. They can be very large (e.g. some aquifers can span a province) or quite small.

Aquifers can have high permeability, so water can flow easily in this layer. There is always a barrier below an aquifer that will not let water pass through to lower layers. This barrier, called an aquitard, is a layer with very low permeability. It is very difficult for water to pass through an aquitard, so it helps contain the water in the aquifer. Every aquifer has an aquitard below it and many also have an aquitard above. When an aquifer has an aquitard on top of it, it is called a confined aquifer. If an aquifer only has an aquitard below it and does not have an aquitard above then it is called an unconfined aquifer. Unconfined and confined aquifers are shown in figure above.

Unconfined aquifers are directly connected to the surface. When water infiltrates into the ground it passes through the unsaturated zone. The unsaturated zone has mostly air in it and is highly permeable. When the water passes the unsaturated zone it reaches the saturated zone. The saturated zone is filled with water. The boundary between the saturated zone and the unsaturated zone is called the water table. When water seeps into the earth it will travel down the soil layers until it reaches the water table. The water table moves up and down depending on how much water is in the unconfined aquifer. Since unconfined aquifers are connected to the surface, they recharge quickly, and are prone to contamination.

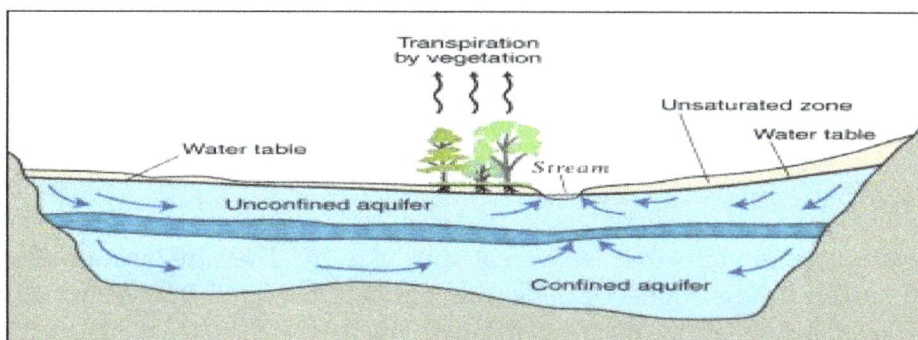

Figure: Confined and unconfined aquifers: This figure shows the two types of aquifers, confined and unconfined, within an impermeable bedrock basin. The dark blue layer between the aquifers is an aquitard.

Confined aquifers are confined because they have an aquitard above and below the aquifer. These aquifers do not recharge quickly because it takes a long time for water to pass through the top aquitard. In some cases, confined aquifers contain high quality water because they are not directly impacted by human activity on the surface. Confined aquifers can contain groundwater that is very old. Water can stay in a confined aquifer for several millennia.

Types of Aquifers

Aquifers are categorized as confined or unconfined, but there are many types of aquifers that are classified by where they are located in the earth and the material of which they are comprised. There are three types of aquifers: *unconsolidated deposit aquifers, bedrock aquifers* and *quaternary aquifers.*

Unconsolidated Deposit Aquifers

An unconsolidated deposit aquifer is an aquifer that is made up of loose sediment such as gravel and sand. These aquifers are close to the surface and are almost always unconfined. This type of aquifer is commonly found near rivers in a floodplain. Unconsolidated deposit aquifers are formed as the result of old rivers that no longer exist, by glaciers that have moved the sediment or by deposition at the bottom of a lake. The water in an unconsolidated deposit aquifer is directly connected to the surface water system.

Bedrock Aquifers

Bedrock is the hard rock that lies below all the sand, gravel and soil near the ground surface. A bedrock aquifer is an aquifer that is confined within hard bedrock layers. Water can travel through porous bedrock, or through cracks, fractures and crevasses in the hard bedrock. In Alberta, 84% of groundwater wells draw from bedrock aquifers. These aquifers are easily accessible in areas where the bedrock is near the earth's surface, such as in southern Alberta.

In Alberta, there are three types of bedrock aquifers: *carbonate aquifers, sandstone aquifers,* and *fractured shale aquifers.*

Carbonate aquifers are made of rocks such as limestone and usually contain saline water. Sandstone aquifers are made of sandstone, a highly permeable rock, and can contain either saline or freshwater. The largest aquifer in Alberta, the Paskapoo Aquifer, is a sandstone aquifer. One third of groundwater wells in Alberta are located in the Paskapoo Aquifer. Shale is a rock that is similar to sandstone, but is less permeable. For shale to be an aquifer, it must be fractured, or cracked, so water can flow into it. Fractured shale aquifers are relatively rare in Alberta. The wells that draw from this type of aquifer do not produce as much water.

Quaternary Aquifers

Quaternary aquifers are aquifers that were created by glaciers. They are located between bedrock and the earth's surface. These aquifers can be confined or unconfined. There are two types of quaternary aquifers: *buried valley aquifers* and *alluvial aquifers.*

Buried valley aquifers are confined aquifers that can be directly above bedrock or higher up in the rock layers. These are ancient valleys that are filled with permeable sand and gravel. Unconfined sand and gravel aquifers are located at the surface or near the surface. An alluvial aquifer is a specific type of unconfined aquifer which has a river flowing through it. The river is the main source of recharge. Quaternary aquifers generally contain freshwater.

Body of Water

Water covers almost 70% of our Earth. The largest body of water is the ocean, while the remaining bodies of water can be subdivided into categories like glaciers and ice caps, groundwater, freshwater, and atmospheric water. In fact, about 97% of our water resources are saltwater, 2% is stored in glaciers and ice caps, and only 1% is freshwater. Of this 1% being freshwater, 97% is in the form of groundwater, often stored in aquifers deep below the soil surface. The rest is stored in rivers and lakes and is more or less directly usable by humans. Only 0.001% of the total water resources are in the atmosphere.

Oceans- The Earth's ocean is the largest and most apparent body of water. All the oceans of the Earth are connected to each other but we often distinguish five distinct oceans through geopolitical rationale: the Pacific Ocean, Atlantic ocean, Indian ocean, Southern ocean, and Arctic ocean. The oceans are a vital part of the Earth ecosystem; they contain the Earth's highest biodiversity, from sharks to phytoplankton, to jellyfish, and not to forget the coral reefs that are themselves living organisms. Oceans determine the wind and climate patterns and are responsible for about 90% of the Earth's natural oxygen production, mainly transpired by phytoplankton. There is still a limitless frontier to explore in the oceans, as half of the world's oceans are over 3000 meters (9,800 ft) deep, and nearly entirely inaccessible. The deepest point is the Marianas Trench in the Pacific Ocean, east of the Philippines and Japan. In 1960 Don Wolsh and Jacques Piccard became the first and only people to reach the bottom of the trench, which has a maximum depth of 10.923 meters (35,838 ft).

Ice- After the oceans, the ice caps and glaciers are the biggest category. This means that an enormous amount of the Earth's water is stored in ice. When one looks at the immense ice sheets of Antarctica, however, it doesn't sound so strange. The Antarctic ice sheet is the largest mass of ice by far and covers an area of almost 14 million km2 (5,400,000 sq mi). In comparison, both Europe and the USA are less than 10 million km2 (3,860,000 sq mi). This ice sheet, including the surrounding frozen sea, contains such a large amount of water that sea levels worldwide would rise more than 60 meters if this would melt. The other two largest ice storages are the North Pole and Greenland. The North Pole is actually a large ice pack floating on the sea. It is two or three meters thick on average. The Greenland ice sheet covers the larger part of Greenland and has an area of 1,755,637 km² (677,676 sq mi). Though smaller than Antarctica it still amasses an area one fifth the size of Europe. Other smaller ice bodies are glaciers and ice caps in the high mountains and in the far North.

Groundwater- A lot of water is stored in the ground. The size of groundwater reserves is often underestimated because water under the ground cannot be seen. Nevertheless, groundwater is the body of water where most of our planet's freshwater is stored, as well as a similar amount of saltwater. The simplest definition of groundwater is "water stored in the ground." Water that infiltrates the soil can fill in the soil pores, the spaces between soil particles. This becomes evident when we dig a hole in the ground, and, at some point or another, we watch as groundwater inevitably starts to fill the bottom of the pit. This is the groundwater table, all the voids and soil pores below this point are saturated with water. The groundwater table is the dividing surface between atmosphere and water. The depth of the groundwater table is often defined in relation to the soil surface, or by NAP, the Amsterdam Ordinance level.

The groundwater table is the upper limit of groundwater. The lower limit is defined as the limit to which water is still circulating through subterranean channels. This is called the hydrological base, or impermeable base. Although there can still be layers of water below this base, this water is no longer considered part of the water cycle, and is therefore often not taken into consideration. Between the water table and the hydrological base there are different soil layers in which groundwater is stored. Some soil layers can contain a lot of water while other layers contain hardly any water. An aquifer is a porous and permeable layer which contains a lot of water. The boundaries of an aquifer are either the surface or other soil layers that are more difficult to permeate. Aquifers where water can flow directly to the surface of the aquifer and vice-versa are called unconfined aquifers. Aquifers that do not have a direct connection with the surface, but are, for instance, trapped under impermeable soil layers, are called confined aquifers. Some of the deepest groundwater aquifers lie hundreds of meters below the surface and contain fossil water, water that has remained in an aquifer for thousands of years.

Freshwater- The abundant fresh surface water bodies on Earth are the centre of human civilizations and rich biodiverse nature systems. The most important surface water bodies are rivers and lakes. Lakes function as water storage facilities and can regulate the sometimes erratic run-off of rivers. Rivers describe water run-off at the surface and are often the product of melting snow and ice in the mountains, supplemented by run-off from rainfall. Rivers can be subdivided into three zones: the production zone, the transport zone, and the deposition zone. The production zone is a network of small streams merging together to form a river. The transport zone is the zone where the river forms one big stream with smaller sub-streams branching to it. In the deposition zone, the river starts to branch again, forming a delta and deposing sand and clay that have been transported from higher altitudes.

The morphology and dynamics of rivers are fascinating, the apparent chaos and randomness of flowing water is immensely well structured when observed close up. One of these structural jewels is the dendrite structure and the related geomorphologic laws of Horton.

- The watercourse of most rivers follows a dendrite structure. This is a transport system that is often used in nature: for instance in trees, in blood vessels, in lightning, and in the lungs. It is the most efficient way of transporting goods. For the sake of scientific conformity, the branches of the dendrite structure are often ordered according to location. The outer branches are labelled as order-1. As soon as two branches of order-1 come together they form a stream of order-2. An order-2 stream that is joined by an order-1 stream remains order-2. However, when two order-2 streams join, they continue as an order-3 stream. This is illustrated in the picture on the right. This structure is irrelative to the size of the system; it is the same for the stream in your backyard as for the currents in the sea.

- The geomorphologic laws of Horton describe the universal traits of dendrite structures. The first geomorphic law of Horton describes the number of branches of a certain order in a dendrite system. It appears that there is a perfect relation between the number of branches in one order, compared to the number of branches in a higher order. This is called the bifurcation number. For example, if the bifurcation number of one river system is 3, and there are nine streams of order-1, it follows that there are three order-2 and one order-1 streams. Worldwide, the range of bifurcation numbers is between 3 and 5, which is surprisingly small for something that seems so random. The bifurcation number for blood vessels, for instance, is 3.4, and for lightning it is 3.2. Apart from the bifurcation number, similar relationships are found in other laws of Horton, e.g. the average length of each order, the average slope, and the average drainage area.

- This is only one example of the structure of river; also the meandering of rivers, sediment transport, drainage area, and river bed profiles are described by a perfect scientific logic.

Atmospheric water- Although the amount of atmospheric water is relatively small, it supplies all the water for surface and groundwater reserves in the form of precipitation. Beside this it also has a protective function: clouds can reduce extreme heat and cold that makes the climate more bearable for living beings. Because precipitation is so crucial to the very sustenance of life itself, we give different names to its various forms. While rain is probably the most common form of precipitation across the globe, other phenomena can be quite fascinating when seen for the first time, such as hail, snow, fog, mist, or dew. In many African countries, snow is a rare phenomenon. The combination of sunlight and atmospheric water furnishes stunning skies and sunsets that have at some time inspired every human being. And on the threshold between rainy and sunlit weather water drops in the air refract the stunning colors of our visual spectrum in the form of a rainbow.

Water Distribution on Earth

Distribution of Earth's Water

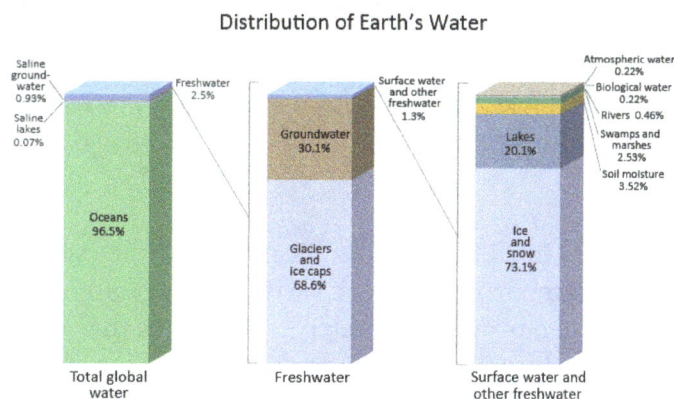

Source: Igor Shiklomanov's chapter "World fresh water resources" in Peter H. Gleick (editor), 1993, Water in Crisis: A Guide to the World's Fresh Water Resources.

Earth's oceans contain 97% of the planet's water, so just 3% is fresh water, water with low concentrations of salts. Most fresh water is trapped as ice in the vast glaciers and ice sheets of Greenland. A storage location for water such as an ocean, glacier, pond, or even the atmosphere is known as a reservoir. A water molecule may pass through a reservoir very quickly or may remain for much longer. The amount of time a molecule stays in a reservoir is known as its residence time. Earth's oceans contain 97% of the planet's water, so just 3% is fresh water, water with low concentrations of salts. Most fresh water is trapped as ice in the vast glaciers and ice sheets of Greenland. A storage location for water such as an ocean, glacier, pond, or even the atmosphere is known as a reservoir. A water molecule may pass through a reservoir very quickly or may remain for much longer. The amount of time a molecule stays in a reservoir is known as its residence time.

Three States of Water

Because of the unique properties of water, water molecules can cycle through almost anywhere on Earth. The water molecule found in your glass of water today could have erupted from a volcano early in Earth history. In the intervening billions of years, the molecule probably spent time in a glacier or far below the ground. The molecule surely was high up in the atmosphere and maybe deep in the belly of a dinosaur. Where will that water molecule go next? Water is the only substance on Earth that is present in all three states of matter – as a solid, liquid or gas. Along with that, Earth is the only planet where water is present in all three states. Because of the ranges in temperature in specific locations around the planet, all three phases may be present in a single location or in a region. The three phases are solid (ice or snow), liquid (water), and gas (water vapor).

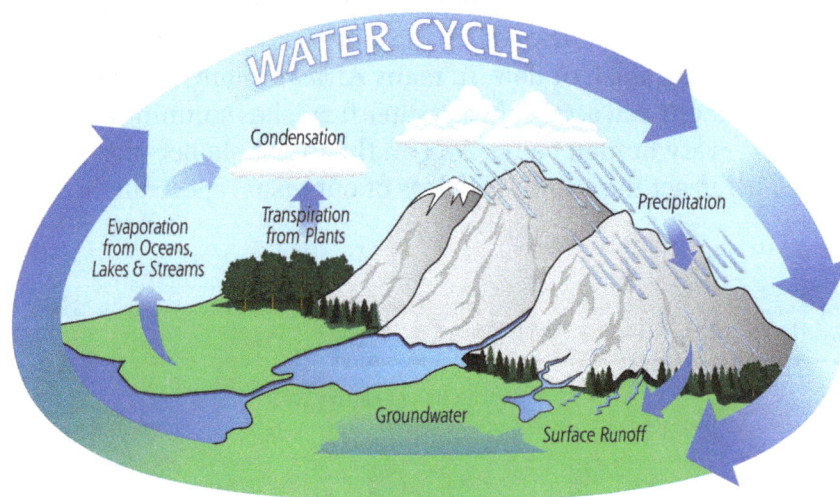

The Water Cycle

Because Earth's water is present in all three states, it can get into a variety of environments around the planet. The movement of water around Earth's surface is the hydrologic (water) cycle. The Sun, many millions of kilometres away, provides the energy that drives the water cycle. Our nearest star directly impacts the water cycle by supplying the energy needed for evaporation. Most of Earth's water is stored in the oceans where it can remain for hundreds or thousands of years.

Water changes from a liquid to a gas by evaporation to become water vapor. The Sun's energy can evaporate water from the ocean surface or from lakes, streams, or puddles on land. Only the water molecules evaporate; the salts remain in the ocean or a freshwater reservoir. The water vapor remains in the atmosphere until it undergoes condensation to become tiny droplets of liquid. The droplets gather in clouds, which are blown about the globe by wind. As the water droplets in the clouds collide and grow, they fall from the sky as precipitation. Precipitation can be rain, sleet, hail, or snow. Sometimes precipitation falls back into the ocean and sometimes it falls onto the land surface.

When water falls from the sky as rain it may enter streams and rivers that flow downward to oceans and lakes. Water that falls as snow may sit on a mountain for several months. Snow may become part of the ice in a glacier, where it may remain for hundreds or thousands of years. Snow and ice may go directly back into the air by sublimation, the process in which a solid changes directly into a gas without first becoming a liquid. Although you probably have not seen water vapor sublimating from a glacier, you may have seen dry ice sublimate in air.

Snow and ice slowly melt over time to become liquid water, which provides a steady flow of fresh water to streams, rivers, and lakes below. A water droplet falling as rain could also become part of a stream or a lake. At the surface, the water may eventually evaporate and re-enter the atmosphere.

A significant amount of water infiltrates into the ground. Soil moisture is an important reservoir for water. Water trapped in soil is important for plants to grow. Water may seep through dirt and rock below the soil through pores infiltrating the ground to go into Earth's groundwater system. Groundwater enters aquifers that may store fresh water for centuries. Alternatively, the water may come to the surface through springs or find its way back to the oceans. Plants and animals depend on water to live and they also play a role in the water cycle. Plants take up water from the soil and release large amounts of water vapor into the air through their leaves, a process known as transpiration. People also depend on water as a natural resource. Not content to get water directly from streams or ponds, humans create canals, aqueducts, dams, and wells to collect water and direct it to where they want it.

Use	United States	Global
Agriculture	34 percent	70 percent
Domestic (drinking, bathing)	12 percent	10 percent
Industry	5 percent	20 percent
Power plant cooling	49 percent	small

The table above displays water use in the United States and globally (Estimated Use of Water in the United States in 2005, USGS). It is important to note that water molecules cycle around. If climate cools and glaciers and ice caps grow, there is less water for the oceans and sea level will fall. The reverse can also happen.

References

- Hydrology: newworldencyclopedia.org, Retrieved 21 May 2018

- Water, science: britannica.com, Retrieved 11 July 2018

- What-is-an-aquifer: albertawater.com, Retrieved 16 May 2018

- Water-bodies-1447: deltawerken.com, Retrieved 31 March 2018

- Distribution-of-earths-water, geophysical: lumenlearning.com, Retrieved 12 May 2018

Hydrological Cycle and Flow

The hydrological cycle, also known as the water cycle, refers to the continuous movement of water between the reservoirs of fresh and saline water, ice and atmospheric water. This cycle is aided by the processes of evaporation, condensation, precipitation, subsurface flow, etc. This chapter closely examines the crucial aspects of hydrological cycle and hydrological flow, such as streamflow, environmental flow, freshwater inflow, water balance, etc.

Hydrological Cycle

The hydrologic cycle is the perpetual movement of water throughout the various components of the Earth's climate system. Water is stored in the oceans, in the atmosphere, as well as on and under the land surface. The transport of water between these reservoirs in various phases plays a central role in the Earth's climate. Water evaporates from the oceans and the land surface into the atmosphere, where it is advected across the face of the Earth in the form of water vapor. Eventually, this water vapor condenses within clouds and precipitates in the forms of rain, snow, sleet, or hail back to the Earth's surface. This precipitation can fall on open bodies of water, be intercepted and transpired by vegetation, and become surface runoff and/or recharge groundwater. Water that infiltrates into the ground surface can percolate into deeper zones to become a part of groundwater storage to eventually reappear as stream-flow or become mixed with saline groundwater in coastal zones. In this final step, water re-enters the ocean from which it will eventually evaporate again, completing the hydrologic cycle. The hydrologic cycle qualitatively, quantitatively, and conceptually is depicted in figures below.

The important reservoirs within the hydrologic cycle include:

Ocean

This vast body of salt water covers 70% of the Earth's surface; it stores and circulates enormous amounts of water and energy. In addition, patterns of ocean surface temperatures can exert a strong influence on circulation patterns in the atmosphere. Frequently, the ocean is divided into two parts, an upper and lower zone. The upper zone is considerably warmer and less saline than the lower zone, and the two are separated by a relatively sharp thermocline. The depth from the ocean surface to the thermocline can be as much as 400 m, but is generally less than 150 m.

Atmosphere

Water can be stored in the atmosphere as liquid in clouds or as water vapor. Water vapor content of the atmosphere is described by its humidity. Specific humidity is a measure of the water content per unit of dry atmosphere (typical values are 1– 20 g kg-1); relative humidity is the amount of water vapor present relative to the amount of water vapour that would saturate the air at a particular temperature. The presence of water in the atmosphere alters the radiation budget of the atmosphere, directly through latent heat and indirectly as both a reflector and absorber of radiation. Water in the atmosphere is the most significant contributor to the natural greenhouse effect.

Cryosphere

The largest stores of fresh water on the Earth are contained in glaciers and icecaps, primarily at high latitudes. The cryosphere has a significant impact on the climate of the Earth because snow and ice-covered surfaces have a very high albedo (comparable to that of clouds). The large volume of runoff from northern high-latitude rivers also influences the Arctic and Atlantic Ocean circulation, which impacts the climate in those regions. Despite the importance of the cryosphere to the hydrologic cycle, relatively little is understood about this part of the climate system, partially because of the lack of adequate data in these often remote and difficult to access areas.

Groundwater

Water beneath the land surface can be classified in a vari-ety of ways. Water closest to the surface (within a few meters) is considered soil moisture, and this water influ-ences the evapotranspiration rate of water from the surface. Soil moisture that is frozen year-round is called permafrost. Deeper below the surface is the aquifer, where the water concentration in the rock and soil is sufficient for withdrawal by pumping. Groundwater for human activities is contained primarily in the aquifer. In this saturated zone, all available spaces within the rock and soil are filled with water. Between the aquifer and soil moisture lays an unsaturated intermediate (vadose) zone that has a lesser influence on the atmosphere than soil moisture. Despite the great societal importance of groundwater supplies, quality spatially distributed subsurface data are elusive.

The Earth System: Physical and Chemical Dimensions of Global Environmental Change

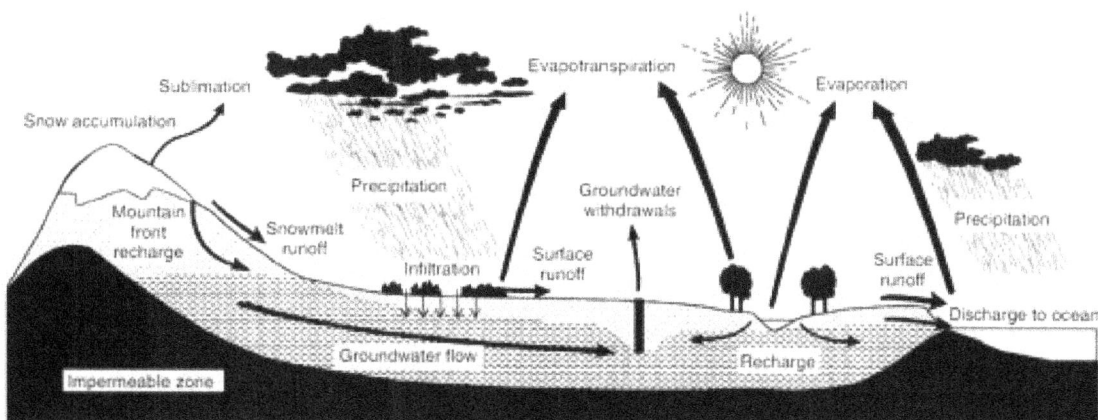

Figure: schematic diagram of various fluxes within the hydrologic cycle.

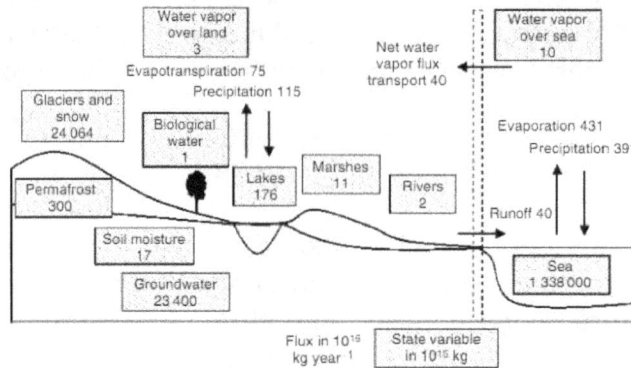

Figure: diagram of the various fluxes and reservoirs within the hydrologic
cycle with their yearly average magnitudes.

Land Surface

Water on land can be contained in lakes and marshes as well as rivers and within living organisms
(biological water). The volume of water stored on land is relatively small, but the flux of water
throughout these systems is relative high. The relevance of this water to human activities para-
mount. In the broadest sense, the major fluxes between reservoirs are:

- Precipitation

Precipitation is the fall of solid or liquid water over land and oceans, and is the major driver of the
hydrologic cycle over land. Hydrologists have traditionally recognized precipitation as the start
of the hydrologic cycle because all other hydrologic phenomena (e.g., evaporation, runoff, and
recharge) result from it. The importance of precipitation to the hydrologic cycle cannot be over-
stated.

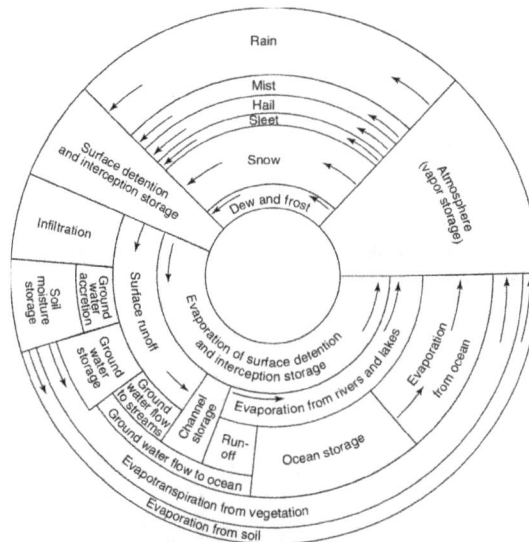

Figure A conceptual diagram of the hydrologic cycle

- Evapotranspiration

Evaporation is the return of water from bare soil or open bodies of water (mainly the ocean surface) to the atmosphere. Transpiration is the transfer of water to the atmosphere through the stomata of vegetation. Collectively, they are considered evapotranspiration.

- Runoff

Runoff is the transport of liquid water across the surface of the Earth. Excess water in saturated soils flows into rivers to the ocean, to terminal lakes or swamps. Groundwater can interact with streamflow in rivers if the water table is near the surface.

- Water vapor transport

Atmospheric water vapor transport is the redistribution of atmospheric water vapor. Globally, there is a net transfer from over ocean to over land. This process is known as advection, and this flux is the major source of water vapor for precipitation over land, aside from recycling.

Descriptions of the Hydrologic Cycle

Mathematical Models

Mathematically, the movement of water throughout the hydrologic cycle can be described using the hydrolog continuity equation:

$$I - O = \frac{\Delta S}{\Delta t}$$

where input (I) and output (O) depend on the reservoir in question (e.g., evapotranspiration is an input to the atmo-sphere, whereas precipitation is an output). The change in storage (S) in time describes the removal from or addition to present supplies to make up for the imbalance between input and output (in the case of the atmosphere, change in storage would signify a change in specific humidity).

The Earth System: Physical and Chemical Dimensions of Global Environmental Change

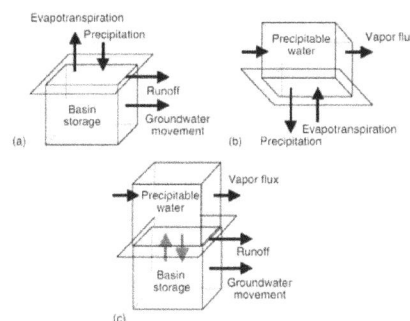

Figure: Mathematical schematic of a water balance for: (a) the land surface;
(b) the atmosphere; and (c) the combined atmosphere and surface.

In contrast to the atmosphere, the water balance of a surface portion of a river basin is considerably more complex. Water is input into this system through precipitation, surface runoff, and groundwater inflow from other parts of the basin. Water is lost through surface runoff, groundwater outflow, and evapotranspiration. The change in storage is reflected in changes in soil moisture content. A graphical depiction of a water balance for a portion of the atmosphere and land surface is shown in figure above.

On a global basis, the Earth is effectively a closed system, and the amount of water present remains relatively constant (i.e., $\Delta S / \Delta t \approx 0$). However, input and output rates of the hydrologic cycle vary regionally and on a wide range of time scales. Describing, quantifying, and predicting these variations are, in essence, major tasks in contemporary hydrology.

Scales for Study of Hydrologic Cycle

From the point of view of hydrologic studies, two scales are readily distinct. These are the global scale and the catchment scale.

Global Scale

From a global perspective, the hydrologic cycle can be considered to be comprised of three major systems; the oceans, the atmosphere, and the landsphere. Precipitation, runoff and evaporation are the principal processes that transmit water from one system to the other. This illustration depicts a global geophysical view of the hydrologic cycle and shows the interactions between the earth (lithosphere), the oceans (hydrosphere), and the atmosphere. The study at the global scale is necessary to understand the global fluxes and global circulation patterns. The results of these studies form important inputs to water resources planning for a national, regional water resources assessment, weather forecasting, and study of climate changes. These results may also form the boundary conditions of small-scale models/applications.

Catchment Scale

While studying the hydrologic cycle on a catchment scale, the spatial coverage can range from a few square km to thousands of square km. The time scale could be a storm lasting for a few hours to a study spanning many years. When the water movement of the earth system is considered, three systems can be recognized: the land (surface) system, the subsurface system, and the aquifer (or geologic) system. When the attention is focused on the hydrologic cycle of the land system, the dominant processes are precipitation, evapotranspiration, infiltration, and surface runoff. The land system itself comprises of three subsystems: vegetation subsystem, structural subsystem and soil subsystem. These subsystems subtract water from precipitation through interception, depression and detention storage. This water is either lost to the atmospheric system or enters subsurface system. The exchange of water among these subsystems takes place through the processes of infiltration, exfiltration, percolation, and capillary rise.

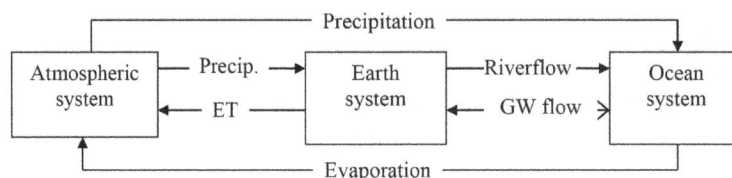

Figure: A global schematic of the hydrologic cycle.

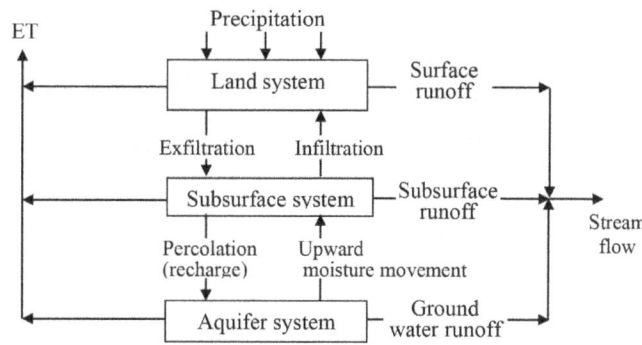

Figure: A schematic of the hydrologic cycle of the earth system.

Time Scales in Hydrologic Cycle

The time required for the movement of water through various components of the hydrologic cycle varies considerably. The velocity of streamflow is much higher compared to the velocity of ground water. The time-step size for an analysis depends upon the purpose of study, the availability of data, and how detailed the study is. The estimated periods of renewal of water resources in water bodies on the earth is given in table below. The time step should be sufficiently small so that the variations in the processes can be captured in sufficient detail but at the same time, it should not put undue burden on data collection and computational efforts.

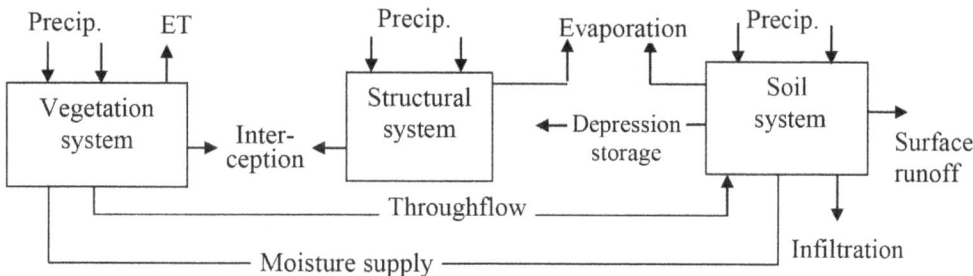

Figure: A detailed schematic of the hydrologic cycle in the land system.

Table: Periods of water resources renewal on the Earth

Water of hydrosphere	Period of renewal
World Ocean	2500 years
Ground water	1400 years
Polar ice	9700 years
Mountain glaciers	1600 years
Ground ice of the permafrost zone	10000 years
Lakes	17 years
Bogs	5 years
Soil moisture	1 year
Channel network	16 days

| Atmospheric moisture | 8 days |
| Biological water | Several hours |

The range of spatial and temporal dimensions of many processes related to the hydrologic cycle.

Humans and The Hydrologic Cycle

One component of the hydrologic cycle that is frequently not directly included in its descriptions is human activity. The hydrologic cycle would continue, irrespective of human activities, but humans do have a significant impact on the terrestrial component of the hydrologic cycle. Likewise, changes in the hydrologic cycle can dramatically impact human activities, for better or for worse. Under growing population pressures, decreasing availability of freshwater per person, and potential global climate change, how this feedback will develop in coming years remains an interesting yet unresolved question.

Are humans affected by the hydrologic cycle? Certainly. Water is essential to life on Earth. The availability of water has shaped where civilizations have developed and thrived, just as lack of water has caused great hardships. Water is both a necessity and a resource for financial gains. Yet, while it is a benefit, it is also a hazard. On average, over $8 billion damages per year have resulted from flooding and hurricanes in the US alone. Indeed, this does not include pandemic health hazards created by poor water quality.

Will humans affect the hydrologic cycle in the future? Undoubtedly. There is considerable evidence that humans have affected the hydrologic cycle in the past. Large dams, reservoirs, and extensive canal systems are perhaps the most visible testimony to this. In several basins across the globe, surface water resources have been so intensively developed that major rivers periodically cease flowing to the ocean (such as the Colorado River in the US and the Yellow River in China). Inter- and intra-basin transfers have significantly interfered with the natural distribution of water.

The most pervasive change to the hydrologic cycle due to human activities is associated with land-use change. These changes are very important to consider as, according to van Dam, "the effects of climate variability and change on the hydrological cycle will be coincident with those of changes in land use, which could be of the same order of magnitude." The various types of land-use changes range from deforestation to agriculture, urbanization to draining swamplands. The impact of these land uses on stream flow is presented in table above. The impacts arise from changes in surface albedo, surface roughness, surface permeability (the ability of water to pass through a surface, such as concrete), and the ability of the surface to intercept and evaporate moisture. These impacts are inherently scale-dependent, and most local land-use change will not have a major impact on the continental and global hydrologic cycle. However, the extent of land-use change in total is considerable; suggested that, over the last 8000 years, approximately 11% of the land surface has been converted to arable land, and 31% of forests have been modified from their original condition. Additionally, certain regions are poised to have a disproportionately strong impact on global circulation. For example, there is debate as to whether Amazon deforestation will have an impact on tropical and extra tropical climate out of the region. Although this is an area of active research, the expected long-range impacts from Amazon deforestation outside the region remain unclear.

While global fluxes or distributions of water may not be influenced by water quality, humans significantly impact water quality at every step of the hydrologic cycle. Since the 1970s, the primary atmospheric water quality concern has been acid rain. Acid rain damages trees, particularly at high elevations, and contributes to the acidification of lakes and streams. Regions already affected include North America and Northern Europe. Contamination of water at and below the land surface poses a significant threat to potable water supplies. Water quality can be affected by human activities in a multitude of ways, including effluent, leeching from landfills, industrial and mining activities, and agricultural fertilizer and pesticide runoff. In particular, the long-term isolation of hazardous radiological by products from water supplies poses a special challenge.

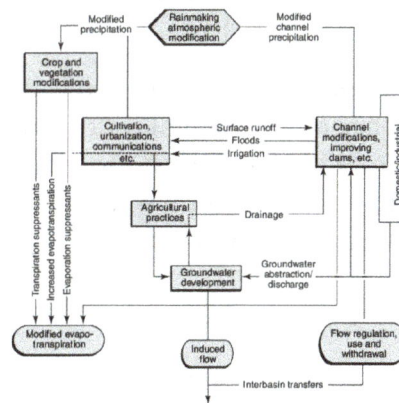

Figure: Systems diagram of the impacts of human activities on streamflow

Precipitation

Precipitation is the falling of water from the sky in different forms. They all form from the clouds which are raised about 8 to 16 kilometers (4 to 11 miles) above the ground in the earth's troposphere. Precipitation takes place whenever any or all forms of water particles fall from these high levels of the atmosphere and reach the earth surface. The drop to the ground is caused by frictional drag and gravity. When one falling particle drops from the cloud, it leaves behind a turbulent wake, causing faster and continued drops.

The (clouds) crystallized ice may reach the ground as ice pellets or snow or may melt and change into raindrops before reaching the surface of the earth depending on the atmospheric temperatures. For this reason, there are many different types of precipitation namely rain, snow, sleet, freezing rain, hail, snow grains and diamond dust. They are forms of water that fall from the sky's frozen clouds.

Rain

Rain is any liquid that drops from the clouds in the sky. Rain is described as water droplets of 0.5 mm or larger. Droplets less than half a millimeter are defined as drizzle. Raindrops frequently fall when small cloud particles strike and bind together, creating bigger drops. As this process continues, the drops get bigger and bigger to an extent where they become too heavy suspend on the air. As a result, the gravity pulls then down to the earth.

When high in the air, the raindrops start falling as ice crystals or snow but melt when as they proceed down the earth through the warmer air. Rainfall rates vary from time to time, for example, light rain ranges from rates of 0.01 to 0.1 inches peer hour, moderate rain from 0.1 to .3 inches per hour, and heavy rain above 0.3 inches per hour. Rain is the most common component of the water cycle and replenishes most of the fresh water on the earth.

Snow

Snow occurs almost every time there is rain. However, snow often melts before it reaches the earth surface. It is precipitation in the form of virga or flakes of ice water falling from the clouds. Snow is normally seen together with high, thin and weak cirrus clouds. Snow can at times fall when the atmospheric temperatures are above freezing, but it mostly occur in sub-freezing air. When the temperatures are above freezing, the snowflakes can partially melt but because of relatively warm temperatures, the evaporation of the particles occurs almost immediately.

This evaporation leads to cooling just around the snowflake and makes it to reach to the ground as snow. Snow has fluffy, white and soft structure and its formation is in different shapes and ways, namely flat plates and thin needles. Each type of snow forms under specific combinations of atmospheric humidity and temperatures. The process of snow precipitation is called snowfall.

Sleet (Ice Pellets)

Sleet takes place in freezing atmospheric conditions. Sleet, also known as ice pellets, form when snow falls into a warm layer then melts into rain and then the rain droplets falls into a freezing layer of air that is cold enough to refreeze the raindrops into ice pellets. Hence, sleet is defined as a form of precipitation composed of small and semitransparent balls of ice. They should not be confused with hailstones as they are smaller in size.

Sleet is often experienced during thunderstorms and is normally accompanied with frosty ice crystals that form white deposits and a mixture of semisolid rain and slushy snow. Ice pellets (sleet) bounce when they hit the ground or any other solid objects and falls with a hard striking sound. Sleet don not freeze into a solid mass except when it combines with freezing rain.

Freezing Rain

Freezing rain happens when rain falls during below freezing conditions/temperatures. This normally results in the solidification of rain droplets. The raindrops are super-cooled while passing through the sub-freezing layer in the atmosphere and freezes by the time it reaches the ground. During freezing rains, it is common to witness an even coating of ice on cars, streets, trees, and power lines. The resulting coating of ice is called glaze and it can build up to a thickness of several centimeters. Freezing rains pose a huge threat to normal operations of roadway transportation, aircrafts, and power lines.

Hail

Hailstones are big balls and irregular lumps of ice that fall from large thunderstorms. Hail is purely a solid precipitation. As opposed to sleets that can form in any weather when there are thunderstorms, hailstones are predominately experienced in the winter or cold weather. Hailstones are mostly made up of water ice and measure between 0.2 inches (5 millimeters) and 6 inches (15 centimeters) in diameter. This ranges in size of a pea's diameter to that larger than a grapefruit.

For this reason, they are highly damaging to crops, tearing leaves apart and reducing their value. Violent thunderstorms with very strong updrafts usually have the capability to hold ice against the gravitational pull, which brings about the hailstones when they eventually escape and fall to the ground. So, hailstones are formed from super-cooled droplets that slowly freeze and results in sheet of clear ice.

Drizzle

Drizzle is very light rain. It is stronger than mist but less than a shower. Mist is a thin fog with condensation near the ground. Fog is made up of ice crystals or cloud water droplets suspended in the air near or at the earth's surface. Drizzle droplets are smaller than 0.5 millimeters (0.02 inches) in diameter. They arise from low stratocumulus clouds. They sometimes evaporate even before reaching the ground due to their minute size. Drizzle can be persistent is cold atmospheric temperatures.

Sun Shower

Sun shower is a precipitation event that is registered when rain falls while the sun shines. It occurs when the winds bearing rain together with rain storms are blown several miles away, thus giving rise to raindrops into an area without clouds. Consequently, sun shower is formed when single rain cloud passes above the earth's surface and the sun's rays penetrate through the raindrops. Most of the time, it is accompanied with the appearance of a rainbow.

Snow Grains

Snow grains are as very small white and opaque grains of ice. Snow grains are fairly flat and have diameter generally less than 1mm. They are almost equivalent to the size of drizzle.

Diamond Dust

Diamond dusts are extremely small ice crystals usually formed at low levels and at temperatures below -30°C. Diamond dust got its name from the sparkling effect which is created when light reflects on the ice crystals in the air.

Interception

Interception deals with the amount of water that is caught and stored on the leaves and stems of vegetation. The term vegetation includes forest tree cover, crops and low level vegetation like under-bush, grass etc. Part of the precipitation which falls either as rainfall or snowfall is caught by vegetative cover before it reaches ground.

This is a matter of common experience when the people take shelter below a tree during rainfall. The rain drops or snowflakes are retained by the leaves as droplets or thin layers on their surface or in the depression of the leaves.

After this storage capacity is exhausted and the rain or snow starts dripping down some water also flows down along the stems of the tree or the plant. The rate of interception is high at the beginning of the rain in case of a dense vegetal canopy during summer. As time progresses and if storm continues the rain starts falling through the tree canopy.

The water caught by the vegetation gets disposed off in three ways namely:

  i.   Through fall;

  ii. Flow along the stem; and

  iii. Evaporation.

Obviously after the storage capacity is exhausted the rate of interception reduces considerably and is equal to that amount which gets evaporated from the vegetation. The amount of precipitation intercepted can be measured by placing several rain-gauges below the vegetal canopy on the ground. Average precipitation that reaches this gauge can be compared with the precipitation measured from a rain-gauge placed in an open area.

The difference between the two gauge readings gives the precipitation intercepted by the vegetation. It may, however, be remembered that all the intercepted water is not lost in evaporation. Depending upon site conditions some water falls down either as a through fall or a stem flow. It is estimated that a dense forest cover intercepts about 10 to 25% of the annual precipitation.

Factors Affecting Interception

The amount of water intercepted is greatly variable and depends on many things. Since the interception affects the distribution of rainfall or snowfall and subsequent run-off, it is necessary to understand the factors which affect interception.

Following are some of the important factors:

  (i) Type of Vegetation:

   Interception varies with the species, its age and density of stands. About 10 to 20% of precipitation occurring in the growing season is intercepted. It is lost substantially by way of evaporation from leaves. In dense tall vegetation interception is quite substantial.

So interception by low-lying vegetation is usually negligible for hydrologic studies. But if such vegetation cover exists below the forest canopy it affects runoff substantially. Dense grasses and herbs approaching full growth intercept as much precipitation as forest cover. However, since their season is short total amount intercepted is considerably less than the forest cover.

(ii) Wind Velocity:

If the wind accompanies the precipitation the leaves become incapable of holding much water as compared with the still air condition. On the other hand due to blowing of wind the evaporation rate also increases. It is thus seen that during short storm interception decreases but during long storm interception is augmented due to increased evaporation.

(iii) Duration of Storm:

Interception storage is filled up in the first part of a rain storm. Therefore, if yearly precipitation is made up of several small duration storms separated by dry spells evaporation will be high and interception will be consequently more. However, if storms of long duration occur and if weather remains cloudy, relatively interception loss will be less.

(iv) Intensity of Storm:

When precipitation occurs in still air conditions with low intensity interception will be more. On the contrary if rain drops come with great speed their impact dislodges intercepted drops and leaves cannot hold much water.

(v) Season of the Year:

During summer or dry season the interception rate is quite high because of high evaporation. Summer interception is 2 to 3 times more than the winter season interception.

(vi) Climate of the Area:

In arid and semiarid regions due to prevailing dry conditions the interception loss is more than that occurring in humid regions.

Groundwater Recharge

Ground water recharge is a vital part of the "hydrologic cycle". The hydrologic cycle is a constant movement of water above, on, and below the earth›s surface. It is a cycle that replenishes ground water supplies. It is not easy to estimate groundwater recharge rates because it is tough to track the amount of water which returns to subsurface water supplies, even though several different techniques can be used to arrive at estimates. It is important to understand how much water is entering a supply of groundwater, as this influences how much water can safely be taken from groundwater supplies for human use.

Estimation Methods

Rates of groundwater recharge are difficult to quantify. Some of the methods are:

- Physical: Physical methods use the principles of soil physics to estimate recharge. The direct physical methods are those that attempt to actually measure the volume of water passing below the root zone. Indirect physical methods rely on the measurement or estimation of soil physical parameters, which along with soil physical principles; can be used to estimate the potential or actual recharge. After months without rain the level of the rivers under humid climate is low and represents solely drained groundwater. Thus the recharge can be calculated from this base flow if the catchment area is known.

- Chemical: Chemical methods utilize the presence of relatively inert water-soluble substances, such as chloride moving through the soil, as deep drainage occurs.

- Numerical models: Recharge can be estimated using numerical methods, using such codes as HELP (Hydrologic Evaluation of Landfill Performance), SHAW, the Shaw Group, WEAP (Water Evaluation And Planning system). These codes generally use climate and soil data to arrive at a recharge estimate.

Importance of Groundwater Recharge

Groundwater recharge is water that has soaked into (infiltrated) the ground, and moved through pores and fractures in soil and rock to the water table. The water table is the depth at which soil and rocks are fully saturated with water. Recharge maintains the supply of fresh water that flows through the groundwater system to wells, streams, springs, and wetlands, which support the plants and animals that are part of the surrounding ecosystem.

In some places (where there are no nearby rivers or large lakes), almost all water-supply needs are met by groundwater, and recharge is critical to maintaining the abundance and quality of groundwater. Groundwater contributes to wells as well as flow to streams, springs, and wetlands year-round, sustaining them during droughts and dry summer months.

Designing Objectives of Groundwater Recharge Structures

In locations where the withdrawal of water is more than the rate of recharge, an imbalance in the groundwater reserves is created. Recharging of aquifers is carried out with the following objectives:

- To maintain natural groundwater as an economic resource.

- To conserve excess surface water underground.

- To combat progressive reduction of groundwater levels.

- To battle unfavorable salt balance and saline water interruption.

- Design of an aquifer recharge system: In order to achieve the objectives, it is crucial to plan out an artificial recharge scheme in a scientific manner. Thus it is very important that a proper scientific examination is carried out for selection of site for artificial recharge of groundwater.

A Proper Groundwater Recharge System Design

- Selection of site: Recharge structures should be planned out after conducting proper hydro-geological investigations. Based on the analysis of this data (already existing or those collected during investigation) it should be possible to establish the maximum rate of recharge that could be achieved at that particular site. This will depend on the quality of soil, presence of underground aquifers, etc.

- Source of water used for recharge: Fundamentally, the potential of rainwater harvesting and the quantity and quality of water available for recharging, have to be assessed.

- Rainwater harvesting: Building structure that can capture rainfall and hold it in place long enough for it to seep through to the underground water table.

Environmental Flow

Rivers, streams and wetlands need certain amounts of water at certain times to support healthy aquatic ecosystems.

In rivers that have been dammed, or are being used for irrigation this normal flow is changed. In other situations, where water is added to a river, such as outflow from a sewage treatment plant, this also alters the natural flow of the river.

To compensate for changes of flow, water may be released from dams or protected from abstraction (this is where water is removed from a river for irrigation or some other purpose), at certain times to allow rivers to function normally. For rivers where more water is being added it may be necessary to allow for more abstraction to regulate the flow.

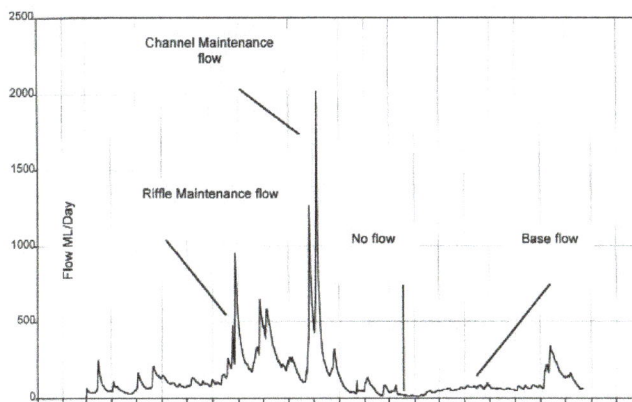

Figure: Seasonal Variation of Flows in the Murrumbidgee River at Lobbs Hole

There are two broad classes of environmental flows: releases of water below dams, and protection of flows in unregulated rivers. Environmental flows are designed to mimic the natural condition of rivers. It is not just about the amount of water but also timing and quality. The Environmental flows are naturally low during summer and autumn, and are much higher during winter and spring. Rivers also naturally experience periods of very low or no flow, and at other times there are floods. It is important that environmental flows mimic this variability of flows. The quality of water released below dams can sometimes be compromised by lower than normal water temperatures, low dissolved oxygen levels, or other water quality parameters. Releasing substandard water quality can severely impair the functioning of aquatic ecosystems.

The Need for Environmental Flows

Streams and rivers need water to survive. Particular flow patterns determine the shape of the stream channel, how different stream habitats connect to each other and what plants and animals occur. For example, fish feed, breed, spawn and migrate in response to natural flow patterns, and may carry out these activities in different parts of stream.

At the same time, it is recognised that water is required for domestic supply, irrigation, and industrial purposes, but taking too much water for these purposes can change the natural flow patterns and can affect habitat availability, food supplies, water chemistry, and nutrient processing. This can lead to the loss of biodiversity, a decline in river water quality, and a decline in the overall condition.

Evolution of Environmental Flow Concepts and Recognition

From the turn of the 20th century through the 1960s, water management in developed nations focused largely on maximizing flood protection, water supplies, and hydropower generation. During the 1970s, the ecological and economic effects of these projects prompted scientists to seek ways to modify dam operations to maintain certain fish species. The initial focus was on determining the minimum flow necessary to preserve an individual species, such as trout, in a river. Environmental flows evolved from this concept of "minimum flows" and, later, "instream flows", which emphasized the need to keep water within waterways.

By the 1990s, scientists came to realize that the biological and social systems supported by rivers are too complicated to be summarized by a single minimum flow requirement. Since the 1990s, restoring and maintaining more comprehensive environmental flows has gained increasing support, as has the capability of scientists and engineers to define these flows to maintain the full spectrum of riverine species, processes and services. Furthermore, implementation has evolved from dam re-operation to an integration of all aspects of water management, including groundwater and surface water diversions and return flows, as well as land use and storm water management. The science to support regional-scale environmental flow determination and management has likewise advanced.

In a global survey of water specialists undertaken in 2003 to gauge perceptions of environmental flow, 88% of the 272 respondents agreed that the concept is essential for sustainably managing water resources and meeting the long-term needs of people. In 2007, the Brisbane Declaration on Environmental Flows was endorsed by more than 750 practitioners from more than 50 countries. The declaration announced an official pledge to work together to protect and restore the world›s rivers and lakes. By 2010, many countries throughout the world had adopted environmental flow policies, although their implementation remains a challenge.

Examples

One effort currently underway to restore environmental flows is the Sustainable Rivers Project, a collaboration between The Nature Conservancy (TNC) and U.S. Army Corps of Engineers (USACE), which is the largest water manager in the United States. Since 2002, TNC and the USACE have been working to define and implement environmental flows by altering the operations of USACE dams in 8 rivers across 12 states. Dam reoperation to release environmental flows, in combination with floodplain restoration, has in some instances increased the water available for hydropower production while reducing flood risk.

Arizona's Bill Williams River, flowing downstream of Alamo Dam, is one of the rivers featured in the Sustainable Rivers Project. Having discussed modifying dam operations since the early 1990s, local stakeholders began to work with TNC and USACE in 2005 to identify specific strategies for improving the ecological health and biodiversity of the river basin downstream from the dam. Scientists compiled the best available information and worked together to define environmental flows for the Bill Williams River. While not all of the recommended environmental flow components could be implemented immediately, the USACE has changed its operations of Alamo Dam to incorporate more natural low flows and controlled floods. Ongoing monitoring is capturing resulting ecological responses such as rejuvenation of native willow-cottonwood forest, suppression of invasive and non-native tamarisk, restoration of more natural densities of beaver dams and associated lotic-lentic habitat, changes in aquatic insect populations, and enhanced groundwater recharge. USACE engineers continue to consult with scientists on a regular basis and use the monitoring results to further refine operations of the dam.

Another case in which stakeholders developed environmental flow recommendations is Honduras' Patuca III Hydropower Project. The Patuca River, the second longest river in Central America, has supported fish populations, nourished crops, and enabled navigation for many indigenous communities, including the Tawahka, Pech, and Miskito Indians, for hundreds of years. To protect the ecological health of the largest undisturbed rainforest north of the Amazon and its inhabitants, TNC and Empresa Nacional de Energía Eléctrica(ENEE, the agency responsible for the project) agreed to study and determine flows necessary to sustain the health of human and natural communities along the river. Due to very limited available data, innovative approaches were developed for estimating flow needs based on experiences and observations of the local people who depend on this nearly pristine river reach.

Methods, Tools and Models

More than 200 methods are used worldwide to prescribe river flows needed to maintain healthy rivers. However, very few of these are comprehensive and holistic, accounting for seasonal and inter-annual flow variation needed to support the whole range of ecosystem services that healthy rivers provide. Such comprehensive approaches include DRIFT (Downstream Response to Imposed Flow Transformation), BBM (Building Block Methodology), and the "Savannah Process" for site-specific environmental flow assessment, and ELOHA (Ecological Limits of Hydrologic Alteration) for regional-scale water resource planning and management. The "best" method, or more likely, methods, for a given situation depends on the amount of resources and data available, the most important issues, and the level of certainty required. To facilitate environmental flow prescriptions, a number of computer models and tools have been developed by groups such as the USACE's Hydrologic Engineering Center to capture flow requirements defined in a workshop setting (e.g., HEC-RPT) or to evaluate the implications of environmental flow implementation (e.g., HEC-ResSim, HEC-RAS, and HEC-EFM). Additionally, a 2D model is developed from a 3D turbulence model based on Smagorinsky large eddy closure to more appropriately model environmental large scale flows. This model is based on a slow manifold of the turbulent Smagorinsky large eddy closure instead of conventional depth-averaging flow equations.

Other tried and tested environmental flow assessment methods include DRIFT, which was recently used in the Kishenganga HPP dispute between Pakistan and India at the International Court of Arbitration.

Subsurface Flow

Part of the infiltrated effective rainfall circulates more or less horizontally in the superior soil layer and appears at the surface through drain channels. This flow is called subsurface flow (in the past it was called hypodermic flow). The presence of a relatively impermeable shallow layer favours this flow. Subsurface flows in water bearing formations have a drainage capacity slower than superficial flows, but faster than groundwater flows. The essential condition for the appearance of the subsurface flows is: the hydraulic lateral conductivity of the environment has to be superior to the vertical conductivity. The subsurface flow in unsaturated regimes can be the base flow in the area with large slopes, and it is dominant in humid regions with vegetal covering and well-drained soils.

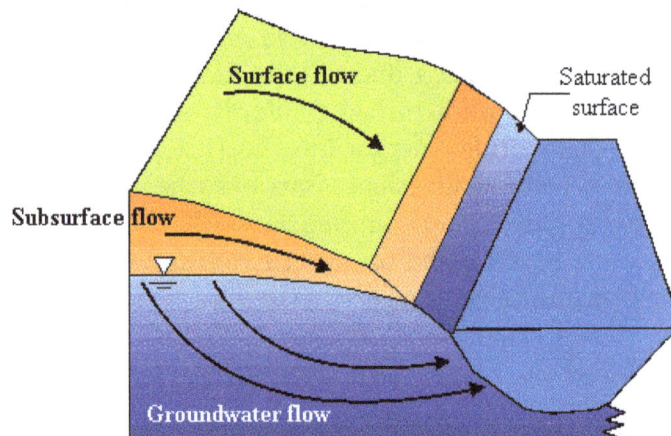

Figure: Different flow types

Piston Effect

Piston effect permits analysis of the subsurface flow. The existence was assumed of a mechanism to transmit a quasi-instantaneous pressure wave. This mechanism, called "the piston effect" assumes that water that falls on a slope is transmitted downstream with a pressure wave and causes a sudden exfiltration on the watershed.

This phenomena principle can be explained by analogy with a saturated soil column to which known quantity of water is added. Due to the gravitation effect water moves to the bottom of the column.

Flow Through Macro Pores

Macro pores are pores in which the capillarity phenomena are inexistent.

The Origin of Macro Pores

We can distinguish:

- Pores formed due to soil micro fauna: with dimension of 1-50 mm. In general they are located in the superior soil layer (0-100 cm).

- Pores formed due to vegetation roots. These pores become free when the plants die. The structure of macro pores network will depend on vegetation type and also on the growing state.

- Natural macro pores: appears when the initial hydraulic conductivity is large.

- Cracks.

Subsurface flow is carried through macro pores. Which lead the water to unsaturated areas.

Pipe Flow

Pipe flow emphasizes soil drainage and water flow. It is difficult to establish the exact difference between pipes and macro pores. Pipes are considered to be larger than macro pores. Also pipes present a higher connectivity degree than pores, without saying that the pipes are forming a continuous network. Pipes lead the water to an unsaturated environment.

Subsurface Stormflow

Subsurface stormflow is a runoff producing mechanism operating in most upland terrains. Subsurface stormflow occurs when water moves laterally down a hillslope through soil layers or permeable bedrock to contribute to the storm hydrograph in a river. In humid environments and steep terrain with conductive soils, subsurface stormflow may be the main mechanism of storm runoff generation. In drier climates and in lowlands with gentler topography, subsurface stormflow may occur only under certain extreme conditions (high rainfall and high antecedent soil moisture), when transient water tables form and induce lateral flow to the channel.

While an important contributor to the volume of flow in the stream, subsurface stormflow is also responsible for the transport of labile nutrients into surface water bodies. Since the flow path of water in the subsurface often determines the chemistry of waters discharging into the stream and hence the water quality, characterizing this subsurface flow path and the water's age and origin is important. Subsurface stormflow may also enhance positive pore pressure development in steep terrain and may be responsible for landslide initiation.

Subsurface stormflow is also known as interflow, lateral flow, subsurface runoff, transient groundwater, or soil water flow.

These multiple terms often confuse the process understanding of subsurface stormflow response to rainfall or snowmelt. While some studies have documented subsurface stormflow as unsaturated flow in the unsaturated zone, most studies have shown that subsurface stormflow is a saturated (or nearsaturated) water flow phenomenon – due either to the rise of an existing water table into more transmissive soil above (with ensuing lateral flow) or the transient saturation above an impeding layer, soil-bedrock interface or some zone of reduced permeability at depth (argillic horizon, hard-pan, plough layer, etc.).

Flow Regimes of Subsurface Stormflow

Subsurface stormflow describes all runoff generation processes in the hillslope close to the soil surface that result in a stream channel hydrograph response during a precipitation or snowmelt event. This response may be coupled directly to flow in preferential pathways like macropores and layers or areas with high permeability. However, rapid subsurface stormflow response may also result from a fast hydraulic response of connected saturated areas in a hillslope in response to infiltrating precipitation. The main flow regimes at the hillslope scale may be subdivided into homogeneous matrix flow and preferential flow.

Homogeneous Matrix Flow

Lateral matrix flow can be a viable subsurface stormflow process if water is already stored in the soil within connected saturated and close-to-saturated areas. These areas may respond quickly to an increase in hydraulic gradient and cross-sectional area due to infiltrating water. This process may occur in slopes where a high-permeable soil layer with high infiltration capacity is situated (parallel) above a low-permeable soil layer (e.g. bedrock, argillic horizon, etc). Since the water in storage on a typical hillslope is large relative to the rainfall depth, this matrix regime often results in a large contribution of pre-event water to the stream as only a small amount of event water is necessary to increase the hydraulic gradient and cross-sectional area in the slope and create a connected transient groundwater body.

Preferential Flow

Preferential lateral flow occurs either in distinctive structure in the soil where water flows only under gravity (macropores) or in areas with a higher permeability than the surrounding soil matrix. Macropores in the soil or fractures in the bedrock that are oriented predominantly slope parallel may transport water efficiently and rapidly from the hillslope to the stream. Laterally oriented macropore flow may dominate in many forest environments where macropores are generated by plant roots and burrowing animals. Macropores that are enlarged by erosion and connected over several meters are often termed soil pipes. If a connected network is developed because of internal erosion and eluviation and connection of macropores, piping can provide effective drainage augmentation to hillslopes. However, disconnected macropores that connect hydraulically during storms can also result in an effective drainage of the hillslopes. If the underlying bedrock is more permeable, water can infiltrate into the bedrock and then percolate vertically into bedrock fissures and cracks negating the macropore enhancement of subsurface stormflow on the timescale of a rainfall event.

High permeability layers are areas in the slope with a coarse texture and large pore and void space. These are often found in talus slopes, landslide debris, peri-glacial solifluction deposits, or unconsolidated moraine material. Erosion of fine sediments by turbulent flow in areas with already coarse soil material increases the hydraulic conductivity and makes these areas particularly conductive. This flow regime can be best envisioned by an extension of the surface streams into the hillslope where many "small" subsurface streams connect preferentially the hillslopes with the streams.

Freshwater Inflow

Freshwater inflow is water that is less saline than marine water, and generally refers to water which flows downstream from the inland sources. This water enters into the bay and mixes with the more saline seawater, creating an estuary area that is less salty than the ocean.

The lower salinity environment created by freshwater inflows is crucial to the productivity of the bay and estuary system. Inflows carry crucial nutrients and sediments into the coastal system and provide the necessary salinity balance that supports nursing and breeding grounds for developing marine life.

Estuaries are ecosystems where freshwater from streams and rivers meet marine waters of coastal bays and mixing occurs. Freshwater inflow is the term referring to the freshwater that flows from these streams and rivers into estuaries. Mixing in estuarine ecosystems occurs spatially and temporally from climatic influences including tidal action, seasonal variability and storms and is affected by the amount of seawater in the estuarine system which is governed by tides. Tides are defined as the periodic rise and fall of the surface of the sea driven by the gravitational pull of the moon and sun. Estuaries are influenced by the tides but are often somewhat protected tidal action (and tropical storms) by buffers further offshore such as barrier islands and peninsulas.

Estuarine ecosystems are among the most productive ecosystems on the planet, and in Texas house such species as blue crab (*Callinectes sapidus*), red drum (*Scianops ocellatus*), southern flounder (*Paralichthys lethostigma*), spotted seatrout (*Cynoscion nebulosus*), and many others. Some economically important estuarine habitats include tidal flats, salt marshes, seagrass beds, oyster reefs, and mangroves.

An estuary cannot properly function without freshwater inflows from rivers and streams. Freshwater inflows are important to estuaries because they provide low-salinity nurseries and transportation of nutrients, sediment, and organic material which effects species movement and reproductive timing.

Global changes largely caused by anthropogenic influences are altering the amount of freshwater inflows to estuaries. Humans are diverting water from rivers and streams, decreasing the amount of flows making it to estuarine ecosystems. As the human population grows and the strain on water resources continues, the ability to effectively manage freshwater inflows into estuaries is becoming a priority worldwide.

Water Balance

The water balance is an accounting of the inputs and outputs of water. The water balance of a place, whether it is an agricultural field, watershed, or continent, can be determined by calculating the input, output, and storage changes of water at the Earth's surface.

Components of Water Balance

Precipitation (P)

Precipitation (also known as one of the classes of hydrometeors, which are atmospheric water phenomena) is any product of the condensation of atmospheric water vapour that is pulled down by gravity and deposited on the Earth's surface. The main forms of precipitation include rain, snow, ice pellets, and graupel. It occurs when the atmosphere, a large gaseous solution, becomes saturated with water vapour and the water condenses, falling out of solution (i.e., precipitates).

Two processes, possibly acting together, can lead to air becoming saturated cooling the air or adding water vapour to the air. Virga is precipitation that begins falling to the earth but evaporates before reaching the surface; it is one of the ways air can become saturated. Precipitation forms via collision with other rain drops or ice crystals within a cloud.

Moisture overriding associated with weather fronts is an overall major method of precipitation production. If enough moisture and upward motion is present, precipitation falls from convective clouds such as cumulonimbus and can organize into narrow rain-bands. Where relatively warm water bodies are present, for example due to water evaporation from lakes, lake-effect snowfall becomes a concern downwind of the warm lakes within the cold cyclonic flow around the backside of extra-tropical cyclones. Lake-effect snowfall can be locally heavy.

Thunder-snow is possible within a cyclone's comma head and within lake effect precipitation bands. In mountainous areas, heavy precipitation is possible where upslope flow is maximized within windward sides of the terrain at elevation. On the leeward side of mountains, desert climates can exist due to the dry air caused by compressional heating. The movement of the monsoon trough, or inter-tropical convergence zone, brings rainy seasons to savannah climes.

Rain drops range in size from oblate, pancake-like shapes for larger drops, to small spheres for smaller drops. Precipitation that reaches the surface of the earth can occur in many different forms, including rain, freezing rain, drizzle, ice needles, snow, ice pellets or sleet, graupel and hail. Hail is formed within cumulonimbus clouds when strong updrafts of air cause the stones to cycle back and forth through the cloud, causing the hailstone to form in layers until it becomes heavy enough to fall from the cloud.

Unlike raindrops, snowflakes grow in a variety of different shapes and patterns, determined by the temperature and humidity characteristics of the air the snowflake moves through on its way to the ground. While snow and ice pellets require temperatures close to the ground to be near or below freezing, hail can occur during much warmer temperature regimes due to the process of its formation. Precipitation may occur on other celestial bodies, e.g., when it gets cold, Mars has precipitation which most likely takes the form of ice needles, rather than rain or snow.

Actual Evapotranspiration (AE)

Evaporation is the phase change from a liquid to a gas releasing water from a wet surface into the air above. Similarly, transpiration is represents a phase change when water is released into the air by plants. Evapotranspiration is the combined transfer of water into the air by evaporation and transpiration. Actual evapotranspiration is the amount of water delivered to the air from these two processes. Actual evapotranspiration is an output of water that is dependent on moisture availability, temperature and humidity.

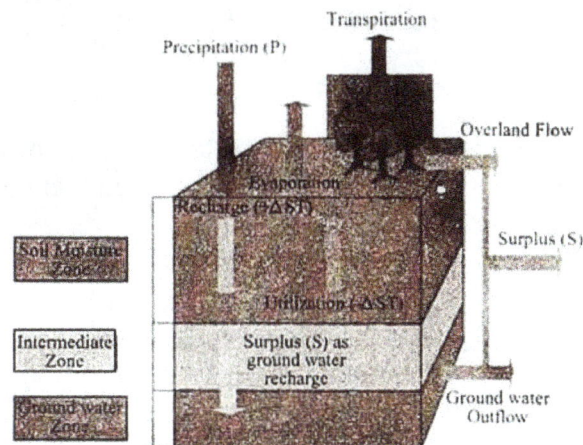

Figure: The soil water balance

Think of actual evapotranspiration as "water use", that is, water that is actually evaporating and transpiring given the environmental conditions of a place. Actual evapotranspiration increases as temperature increases, as long as there is water to evaporate and for plants to transpire. The amount of evapotranspiration also depends on how much water is available, which depends on the field capacity of soils. In other words, if there is no water, no evaporation or transpiration can occur.

Potential Evapotranspiration (PE)

The environmental conditions at a place create a demand for water. Especially in the case for plants, as energy input increases, so does the demand for water to maintain life processes. If this demand is not met, serious consequences can occur. If the demand for water far exceeds that which is actual present, dry soil moisture conditions prevail. Natural ecosystems have adapted to the demands placed on water.

Potential evapotranspiration is the amount of water that would be evaporated under an optimal set of conditions, among which is an unlimited supply of water. Think of potential evapotranspiration of "water need". In other words, it would be the water needed for evaporation and transpiration given the local environmental conditions. One of the most important factors that determine water demand is solar radiation.

As energy input increases the demand for water, especially from plants increases. Regardless if there is, or isn't, any water in the soil, a plant still demands water. If it doesn't have access to water, the plant will likely wither and die.

Soil Moisture Storage (ST)

Soil moisture storage refers to the amount of water held in the soil at any particular time. The amount of water in the soil depends on soil properties like soil texture and organic matter content. The maximum amount of water the soil can hold is called the field capacity. Fine grain soils have larger field capacities than coarse grain (sandy) soils. Thus, more water is available for actual evapotranspiration from fine soils than coarse soils. The upper limit of soil moisture storage is the field capacity, the lower limit is 0 when the soil has dried out.

Change in Soil Moisture Storage (ÄST)

The change in soil moisture storage is the amount of water that is being added to or removed from what is stored. The change in soil moisture storage falls between 0 and the field capacity.

- Deficit (D)

  A soil moisture deficit occurs when the demand for water exceeds that which is actually available. In other words, deficits occur when potential evapotranspiration exceeds actual evapotranspiration (PE > AE). Recalling that PE is water demand and AE is actual water use (which depends on how much water is really available), if we demand more than we have available we will experience a deficit. But, deficits only occur when the soil is completely dried out. That is, soil moisture storage (ST) must be 0. By knowing the amount of deficit, one can determine how much water is needed from irrigation sources.

- Surplus (S)

  Surplus water occurs when P exceeds PE and the soil is at its field capacity (saturated). That is, we have more water than we actually need to use given the environmental conditions at a place. The surplus water cannot be added to the soil because the soil is at its field capacity so it runs off the surface. Surplus runoff often ends up in nearby streams causing stream discharge to increase. Knowledge of surplus runoff can help forecast potential flooding of nearby streams.

- Surface Runoff

  Surface runoff is the water flow that occurs when soil is infiltrated to full capacity and excess water from rain, snowmelt, or other sources flows over the land. This is a major component of the hydrologic cycle. Runoff that occurs on surfaces before reaching a channel is also called a nonpoint source. If a nonpoint source contains man-made contaminants, the runoff is called nonpoint source pollution.

  A land area which produces runoff that drains to a common point is called a watershed. When runoff flows along the ground, it can pick up soil contaminants such as petroleum, pesticides (in particular herbicides and insecticides), or fertilizers that become discharge or non-point source pollution.

  Surface runoff can be generated either by rain fall or by the melting of snow, ice, or glaciers. Snow and glacier melt occur only in areas cold enough for these to form permanently. Typically snowmelt will peak in the spring and glacier melt in the summer, leading to pronounced flow maxima in rivers affected by them. The determining factor of the rate of melting of snow or glaciers is both air temperature and the duration of sunlight. In high mountain regions, streams frequently rise on sunny days and fall on cloudy ones for this reason.

Type of Surface Water Balance

The water balance of the entire mine, a number of components, or a single entity, such as the heap leach pad, may be quantified as part of the water quality and/or quantity management activities at a mine site.

Reasons for undertaking a facility or site water balance study may include:

(a) Evaluate strategies for optimum use of limited water supplies;

(b) Establish procedures for limiting site discharge and complying with discharge requirements, particularly control of the quality of the water and/or the quantity of contaminants discharged from the site; and

(c) Limiting or controlling erosion due to flow over exposed surfaces or in channels, swales, and creeks; and

(d) Estimating the demands on water treatment plants, holding ponds, evaporation ponds, or wetlands.

Analytical Approaches

The most common way to build a water balance model of a facility of site is to use that famous stand-by, Excel. The reason is that most water balance models generally involve no more than successive solution for each component of a facility and hence for each facility of the simple equation
−Inflow − Outflow = Change in Storage

Use of the Mine Water Balance Model

The following are the steps in setting up, refining, and using a water balance model of your mine:

i. Model:

    Have an effective, robust, calibrated and easily updated and adjusted water quality and quantity (volumetric flow) model to understand the complex relationships of the mine for the prediction of water changes. Model all sources of contamination and the inputs as well as outputs. The stakeholders should agree that the model is accurate and appropriate.

ii. Measure:

    Have an effective sampling program to keep the water quality and quantity model up to date and continuously evaluate its effectiveness and test its assumptions.

iii. Calibrate:

    Have the model checked each week initially (then monthly) with water quality and quantity numbers and monitor discrepancies between reality and model; evaluate and explain discrepancies.

iv. Contingency plans:

Have a full set of costed contingency plans with established implementation timelines. Complete the engineering for likely long-term options.

v. Manage:

Understand all possible actions that can be taken to minimise water quality and quantity issues and have them costed to +/− 35% accuracy. Know at what levels what actions need to be taken when pre-specified levels are reached so that management can confidently make decisions which meet its license limits while incurring the least expenditure.

General Relevant Information

Here is a brief description of the various types of data you need to model the water balance of your mine.

Climate

Data are required to quantify – precipitation, snow depths and melt patterns, evaporation, evapotranspiration, wind, and solar radiation. Such data may come from one or more of many sources, including site measurement records, regional databases—usually the local airport, local and national databases accessible on the web, or synthetically generated data that many computer codes produce.

Surface Water

Data may be needed for a water balance study about local stream flow, surface runoff patterns and quantities, and infiltration patterns and rates. Establishing these quantities may again involve consulting site-specific measurement records, local data bases, or running computer codes that enable one to calculated infiltration through a soil surface such as the cover of a waste pile. The most common code for infiltration estimation is HELP, copies of which are commercially available from many vendors.

Groundwater

Groundwater flow patterns and rates must be known or predicted to model the water balance of a facility or mine. At some sites, the groundwater emerges as springs which add to the quantity (as sometimes the constituent loading) of a site. At most facilities and mines, protection of groundwater quality by limiting seepage to the groundwater is a prime objective. Quantification of groundwater flow regimes is complex even at the simplest of sites, and usually involves detailed site-specific studies based on monitoring wells and a history of water quality sampling.

Facility Layout

Before starting a water balance study it is imperative that you have good information about the site and facility layout. This includes – quantification of area, topography, runoff, slopes, location and condition of streams and man-made channels, and possibly even the layout of the mine pit itself.

Preferable the data should include digital maps that may be used with CADD systems to calculate areas, slopes, etc.

Facility Material Characteristics

Geologist and geotechnical engineers will probably have to be involved to characterise the materials of the facilities that are part of the water balance study. The prime characteristic is, or more correctly the hydraulic conductivity, of the soils and rocks that make up the strata at the site, that constitute the mass of the waste rock dump, heap leach pad, or tailings impoundment, or which serve as the cover of reclaimed and closed waste piles. Sampling and laboratory testing quantify the hydraulic conductivity of soil and rock. In situ wells testing quantifies bedrock permeability.

Vegetation

Evapotranspiration via vegetation is often the primary route by which water is lost or removed from a mine water balance system. The analyst needs to know the types and distribution of vegetation. Most computer codes that enable the analyst to quantify evapotranspiration require input of vegetation coverage, density, rooting depth, and periods of growth and quiescence. Collect such information by field observation, and supplement with studies, in situ testing, regional studies, or calibration of models by collecting data and comparing measured and calculated quantities.

Streamflow

Stream flow or discharge is the volume of water that moves through a specific point in a stream during a given period of time. Discharge is usually measured in units of cubic feet per second (cfs). To determine discharge, a cross-sectional area of the stream or river is measured. Then, the velocity of the stream is measured using a Flow Rate Sensor. The discharge can then be calculated by multiplying the cross-sectional area by the flow velocity.

Stream flow is an important factor in the stream ecosystem and is responsible for many of the physical characteristics of a stream. Stream flow can also modify the chemical and biological aspects of a stream. Aquatic plants and animals depend upon stream flow to bring vital food and nutrients from upstream, or remove wastes downstream.

Stream flow has two components. The first is flow velocity, and the second is the volume of water in the stream.

Flow velocity is influenced by the slope of the surrounding terrain, the depth of the stream, the width of the stream, and the roughness of the substrate or stream bottom. If the surrounding terrain is steep, then rain water and snow melt will have less time to soak into the ground and runoff will be greater. In an area with level terrain, such as farm land, the rain water has plenty of time to soak into the ground and there is less runoff. The flow velocity will also vary as the width or depth of a stream changes. For instance, if you squeeze a water hose with your hand, the flow velocity of the water increases. This is because you have reduced the area that the water must flow through, while the volume of water passing through the hose remained constant. The same thing happens in a stream when the stream channel changes in its width or depth. The substrate of the stream bottom also affects the flow velocity since water moves faster over a smooth surface than a rough surface. Flow velocity is greater when the stream bottom is comprised of sand and clay and lower when it is cobble, rock, and boulders.

The volume of water in the stream is affected by the climate of the region. Areas with more rain and snow will have more water draining into surrounding streams and rivers. Seasonal changes affect stream volume as well. In the summer there will be less water in the stream compared to the winter. The number of tributaries that merge with a stream or river contribute more water to the system, increasing the stream volume. Humans are also responsible for altering the volume of water in streams. Water is removed for consumption, industry, and irrigation. Roads and parking lots cover vast areas, preventing rain-water from soaking into the ground. Instead, the water is forced to run off into surrounding streams and rivers.

References

- Hydrologic-cycle: researchgate.net, Retrieved 28 May 2018
- Different-types-of-precipitation, geography: eartheclipse.com, Retrieved 09 July 2018
- Interception-concept-and-factors-affecting-it-60454: yourarticlelibrary.com, Retrieved 27 June 2018
- What-groundwater-recharge, earth-matters: innovateus.net, Retrieved 16 May 2018
- Environmental-flows-fact-575687: environment.act.gov.au, Retrieved 22 April 2018
- Water-balance-meaning-components-and-types-hydrology-geology-5241: geographynotes.com, Retrieved 19 March 2018

Hydrogeology: An Introduction

Hydrogeology is an area of geology which deals with the movement and distribution of groundwater in the soil and rocks of the Earth's surface. All the important aspects of hydrogeology such as groundwater flow and vadose zone have been extensively covered in this chapter.

Hydrogeology is a branch of geology, the study of rocks and the structures that are formed over past periods of time. The way that sediments have been deposited to create different layers of rock beneath the surface, or the way that rocks have been heated and folded over millions of years to create complex structures are the subjects of geology.

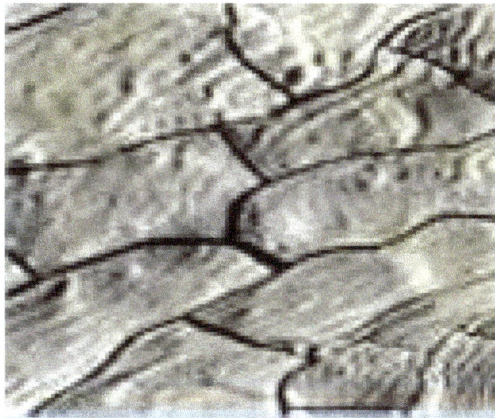

Hydrogeology looks at how water interacts with geological systems. But there is more to hydrogeology than wet rocks. Water is a vital natural resource for people all around the world - whether it is piped to homes or drawn out of wells. Understanding where it is and how it moves under the ground is essential in protecting this resource.

By using geological maps and taking samples of underground and surface water for analysis, hydrogeologists are able build detailed pictures of how water flows through porous rocks underground.

Importance of Hydrogeology

Some parts of the world are blessed with frequent rainfall and plentiful surface water resources, but most countries need to use the water that is stored underground to supplement their needs. Hydrogeologists can help by locating suitable reserves and by assessing how much water it is possible to extract without permanently damaging underground aquifers or surrounding ecosystems.

By looking at the way that groundwater flows, hydrogeologists can also identify when it is at risk from pollution and how we can protect it by careful planning and land-use. Without the knowledge that hydrogeology gives us we have no way of predicting what will happen to our water resources or the future effects on the environment of water extraction.

Groundwater - A Hidden Resource

When we think of fresh water we probably think of rivers, lakes and perhaps reservoirs. Yet even when we consider the huge lakes of North America or mighty rivers like the Amazon and the Ganges, over 97% of the planet's fresh water is to be found under the surface of the earth in the form of groundwater.

In many parts of the world groundwater is the main source of water for day to day use. Boreholes driven down into the saturated layer under the ground yield water for irrigation and urban use. At least 1,500 million urban dwellers are supplied with water from groundwater reserves. It is also extensively used as a low cost rural water supply.

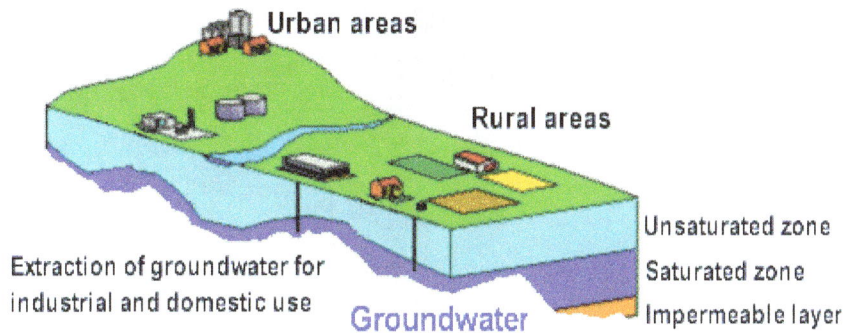

Groundwater has many benefits. It is cheap to develop because of its naturally good quality and widespread occurrence. It is reliable even in times of drought due to the large amounts stored under the surface. It is also protected against catastrophic events - if natural disasters or war disrupt surface water distribution, then groundwater reserves can easily be developed.

Some areas near the surface may not be saturated with water and only have an intermittent supply of water. But, many areas deeper under the surface are saturated with water accumulated over hundreds or even thousands of years. These areas often supply permanent springs and can be tapped with boreholes to provide water. When we draw groundwater from deep aquifer supplies, we are tapping into water locked away under the surface and filtered through layer upon layer of rock. This is one of the reasons why groundwater is so reliable. The water in these areas may even be protected against recent surface pollution, providing a high quality supply where none was previously available.

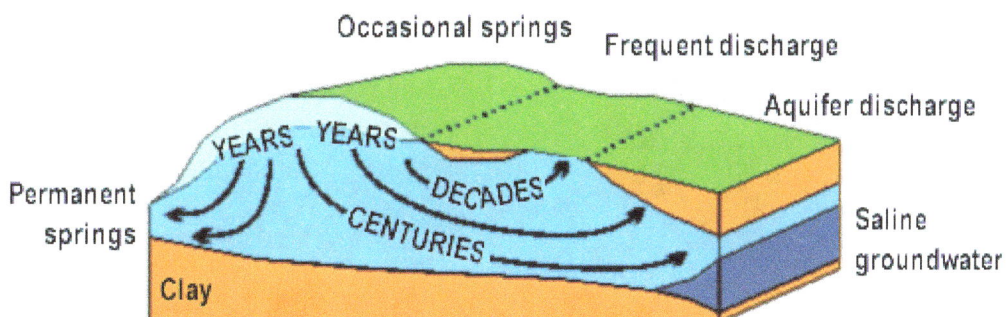

Pollution

In recent years we have become more aware of environmental threats than we once were. Many countries, especially more economically developed nations, have laws and guidelines to help protect the environment. Economic development brings industry and larger cities. Both demand more services and amenities. This places greated pressure on the environment and there are many possible sources of pollution.

Yet surely our water is safe? In developed nations water is brought to our homes in sealed pipe systems kept well away from sources of pollution. Even in developing nations the dangers of pollutants seeping through the surface into wells must be small.

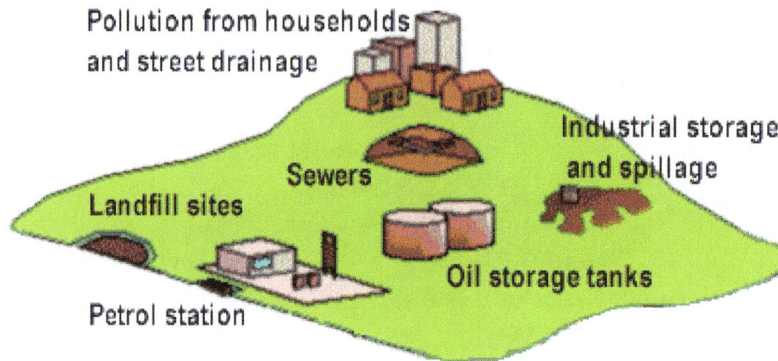

A Fragile Resource

Groundwater is vulnerable to pollution. Some of the features of groundwater that make it so useful also make it a fragile resource. The way that groundwater moves under the surface of the earth means that even seemingly well separated waste and fresh water systems can come into contact.

Leaks from rusting storage tanks in a closed down petrol station might eventually enter public water supplies, even if extraction takes place many kilometres away from the original source of pollution. Without accurate hydrogeological information on groundwater movement it is impossible to plan development and extraction in a sustainable and safe way.

It is not just urban areas that contribute to possible pollution. With time, pesticides or fertilisers used in agriculture can enter groundwater systems, and even manure spreading and ploughing can have an impact.

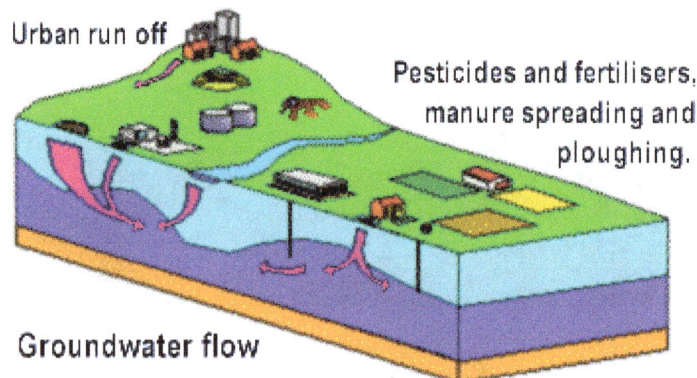

Protecting Groundwater

Protecting groundwater involves deciding how much it possible to extract without risking permanent deterioration in quality. Usually this means not exploiting reserves more quickly than rainfall replenishes them.

It also means thinking about the surrounding environment. Streams and wetlands in may rely on groundwater and be damaged by over-extraction. Possible sources of pollution and how they are likely to interact with groundwater must also be considered.

- Evaluation of groundwater resources.
- Defining optimal exploitation strategies.
- Taking account of side-effects of exploitation.
- Diagnosing vulnerability to quality deterioration.

The Role of Hydrogeologists

By mapping the way that groundwater flows, hydrogeologists are able to identify where groundwater may be vulnerable and where to place potential sources of pollution to minimise risk:

IAH members have been instrumental in developing techniques for investigating groundwater flow.

Working groups of the IAH have helped to standardise hydrogeological mapping and share information internationally.

A major achievement of the IAH has been to make a complete hydrogeological map of Europe.

Hydrogeologists are able to give information on groundwater availability and vulnerability. The best way to conserve the quality of groundwater is to adopt the concept of pollution prevention. This protects the whole resource and avoids the need for treatment at the point of use. Expert advice from hydrogeologists helps to develop codes of practice for land-use and planning, and make this concept a reality.

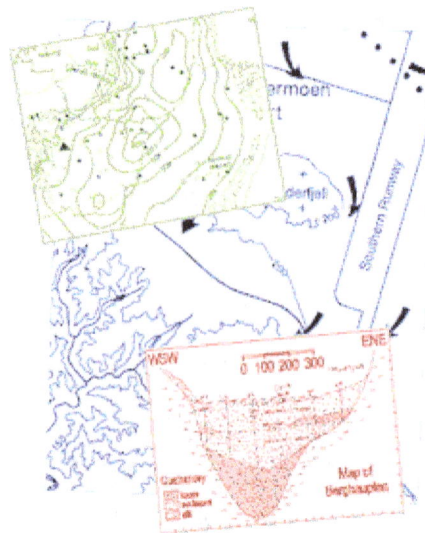

Groundwater

Groundwater or ground water is water located within the ground›s zone of saturation, where the soil pore spaces and fractures in the rock are completely filled with water. It differs from soil water, which is the water that is in the unsaturated zone, or zone of aeration, where the soil pore spaces contain air and water but are not completely saturated. The term groundwater also has been used more broadly as any water beneath the earth's surface in soil; however, the above definition is aligned with those provided by such sources as the U.S. Geological Survey, New York State Dept. of Environmental Conservation, and Pennsylvania Groundwater Policy Education Project.

The depth at which soil pore spaces or fractures and voids in rock become completely saturated with water is called the water table; in other words, below the level of the water table the soil pores and rock fractures are saturated with water. An aquifer is a layer within the zone of saturation that can readily yield and store water, such as in interconnected spaces (fractures, cracks, poor spaces, etc.) that can provide a source of water for a well.

As part of the hydrologic cycle, groundwater stores and transmits water that has filtered down from the surface and it also slowly flows back to the surface, with natural discharge at places such as springs, seeps, and wetlands. Groundwater discharging into a stream provides water to allow the stream to flow throughout the year. Groundwater also is withdrawn for agricultural, municipal, and industrial use by constructing and operating extraction wells.

Although a vitally important renewable resource, that serves many critical economic and environmental needs, groundwater reserves in various regions face such threats as depletion from overdraft and contamination.

Storage of Groundwater

Groundwater is stored in the tiny open spaces between rock and sand, soil, and gravel. How well loosely arranged rock (such as sand and gravel) holds water depends on the size of the rock particles. Layers of loosely arranged particles of uniform size (such as sand) tend to hold more water than layers of rock with materials of different sizes. This is because smaller rock materials settle in the spaces between larger rock materials, decreasing the amount of open space that can hold water. Porosity (how well rock material holds water) is also affected by the shape of rock particles. Round particles will pack more tightly than particles with sharp edges. Material with angular-shaped edges has more open space and can hold more water.

Groundwater is found in two zones. The unsaturated zone, immediately below the land surface, contains water and air in the open spaces, or pores. The saturated zone, a zone in which all the pores and rock fractures are filled with water, underlies the unsaturated zone. The top of the saturated zone is called the water table. The water table may be just below or hundreds of feet below the land surface.

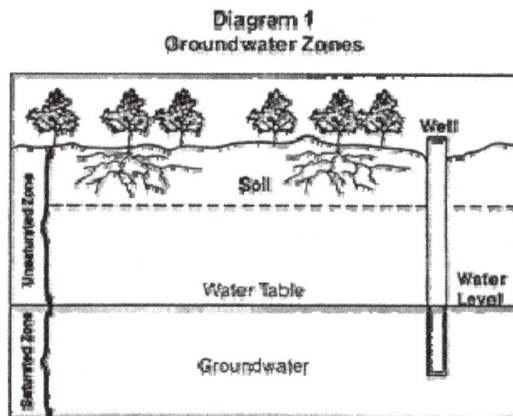

Figure: Groundwater zones

Aquifers

An *aquifer* is an underground geological formation in the zone of saturation that consists of a layer of porous substrate that can readily contain and yield groundwater.

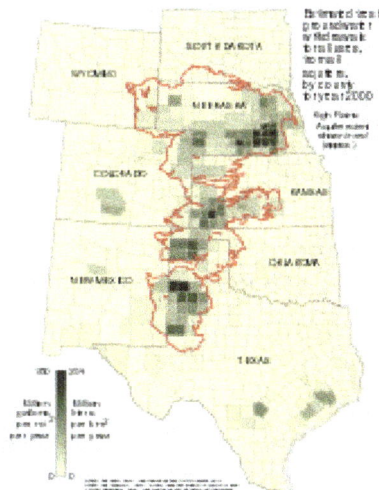

Groundwater withdrawal rates from the Ogallala Aquifer in the central U.S.

Aquifers can be classified as unconfined aquifers and confined aquifers. An unconfined aquifer is one whereby the water table is at or near atmospheric pressure and the water can flow directly to the surface. A well penetrating an unconfined aquifer would have the same water level as the water table outside the well. A confined aquifer or artesian aquifer is one whereby the groundwater is bounded within layers of impermeable substances like dense rock or clay and is very often under pressure. As such, if a well were to be tapped into a confined aquifer, the artesian pressure would force the water to rise in the well to a level higher than the water table, including sometimes above the land surface, as with an artesian well, where the water flows without the need of a pump.

Aquifers also can be classified as consolidated aquifers and unconsolidated aquifers. A consolidated aquifer holds water in interconnected spaces between rock layers, fractures, small cracks, pore spaces, and/or solution channel openings. Limestone, granite, and sandstone are some of the rock types with consolidated aquifers. Limestone aquifers in particular can hold and yield substantial amounts of water, whereas granite small amounts and sandstone moderate amounts. An unconsolidated aquifer involves rock debris or weathered bedrock where soil particles hold water in spaces between the particles. Clay and silt may hold a lot of water but release it very slowly, while coarse-grained sand and gravel may hold less water but release it more freely.

Aquifers also can be classified as consolidated aquifers and unconsolidated aquifers. A consolidated aquifer holds water in interconnected spaces between rock layers, fractures, small cracks, pore spaces, and/or solution channel openings. Limestone, granite, and sandstone are some of the rock types with consolidated aquifers. Limestone aquifers in particular can hold and yield substantial amounts of water, whereas granite small amounts, and sandstone moderate amounts. An unconsolidated aquifer involves rock debris or weathered bedrock where soil particles hold water in spaces between the particles. Clay and silt may hold a lot of water but release it very slowly, while coarse-grained sand and gravel may hold less water but release it more freely.

There may be several diverse aquifers within the zone of saturation, separated by geological formations called aquitards. Aquitards are layers that resist the flow of water from one aquifer to another, such as with nonporous rock or clay with tiny, poorly connected pores. An *aquiclude* is a substrate with porosity that is so low it is virtually impermeable to groundwater.

The characteristics of aquifers vary with the geology and structure of the substrate and topography in which they occur. In general, the more productive aquifers occur in sedimentary geologic formations. By comparison, weathered and fractured crystalline rocks yield smaller quantities of groundwater in many environments. Unconsolidated to poorly cemented alluvial materials that have accumulated as valley-filling sediments in major river valleys and geologically subsiding structural basins are included among the most productive sources of groundwater.

The high specific heat capacity of water and the insulating effect of soil and rock can mitigate the effects of climate and maintain groundwater at a relatively steady temperature. In some places where groundwater temperatures are maintained by this effect at about 10°C (50°F), groundwater can be used for controlling the temperature inside structures at the surface. For example, during hot weather relatively cool groundwater can be pumped through radiators in a home and then returned to the ground in another well. During cold seasons, because it is relatively warm, the water can be used in the same way as a source of heat for heat pumps that is much more efficient than using air.

Circulation of Water

Surface water and groundwater are part of the hydrologic cycle, the constant movement of water above, on, and below the earth's surface. The cycle has no beginning and no end, but you can understand it best by tracing it from precipitation.

Precipitation occurs in several forms, including rain, snow, and hail. Rain, for example, wets the ground surface. As more rain falls, water begins to filter into the ground. How fast water soaks into, or infiltrates the soil depends on soil type, land use, and the intensity and length of the storm. Water infiltrates faster into soils that are mostly sand than into those that are mostly clay or silt. Almost no water filters into paved areas. Rain that cannot be absorbed into the ground collects on the surface, forming runoff streams.

When the soil is completely saturated, additional water moves slowly down through the unsaturated zone to the saturated zone, replenishing or recharging the groundwater. Water then moves through the saturated zone to groundwater discharge areas.

Evaporation occurs when water from such surfaces as oceans, rivers, and ice is converted to vapor. Evaporation, together with transpiration from plants, rises above the Earth's surface, condenses, and forms clouds. Water from both runoff and from groundwater discharge moves toward streams and rivers and may eventually reach the ocean. Oceans are the largest surface water bodies that contribute to evaporation.

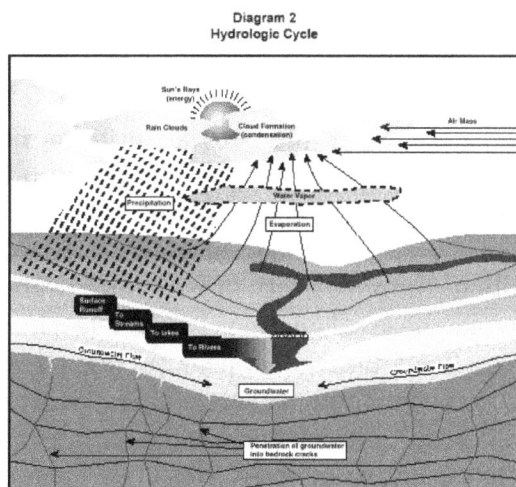

Figure: Hydrologic Cycle

Issues

Two key issues facing groundwater reserves are (1) depletion of groundwater; and (2) contamination.

Groundwater is depleted as is pumped out and used faster than it is replenished. This can have the effect of lowering the water table, which in tern can cause drying up of wells and the need for a well owner to deepen the well, lower the pump, or drill a new well, and greater energy costs for operation a pump; reduction of water that goes back into streams and lakes and loss of wildlife habitat and vegetation; and land subsidence. This last issue can arise when the loss of water causes soil to compact, collapse,and drop, and thus the loss of support below ground for structures on the surface.

Groundwater contamination can occur from a number of sources. Toxins can filter down and waste from landfills and agricultural runoff. As water tables are lowered, saltwater contamination can increase, as the freshwater/saltwater boundary is disrupted and saltwater migrates inward as well as upward from the saline groundwater.

Furthermore, as water moves through the landscape, it collects soluble salts, mainly sodium chloride. As the water enters the atmosphere through evapotranspiration, these salts are left behind. In irrigation districts, poor drainage of soils and surface aquifers can result in water tables' coming to the surface in low-lying areas. Major land degradation problems of soil salinity and waterlogging result, combined with increasing levels of salt in surface waters. As a consequence, major damage has occurred to local economies and environments.

Unlike river waters being overused and polluted, groundwater problems are less evident, as aquifers are out of sight. Another problem is that water management agencies, when calculating the "sustainable yield" of aquifer and river water, have often counted the same water twice, once in the aquifer, and once in its connected river. This problem, although understood for centuries, has persisted, partly through inertia within government agencies.

In general, the time lags inherent in the dynamic response of groundwater to development have been ignored by water management agencies, decades after scientific understanding of the issue was consolidated. In brief, the effects of groundwater overdraft (although undeniably real) may take decades or centuries to manifest themselves. In a classic study in 1982, Bredehoeft and colleagues modeled a situation where groundwater extraction in an intermontane basin withdrew the entire annual recharge, leaving "nothing" for the natural groundwater-dependent vegetation community. Even when the borefield was situated close to the vegetation, 30% of the original vegetation demand could still be met by the lag inherent in the system after 100 years. By year 500, this had reduced to 0%, signalling complete death of the groundwater-dependent vegetation. The science has been available to make these calculations for decades; however, in general water management agencies have ignored effects that will appear outside the rough time frame of political elections. Sophocleous (2002) argues that management agencies must define and use appropriate time frames in groundwater planning. This will mean calculating groundwater withdrawal permits based on predicted effects decades, sometimes centuries in the future.

Overdraft

Over-use of groundwater, known as overdraft, can lead to depletion and cause major problems to human users and to the environment. The most evident problem (as far as human groundwater use is concerned) is a lowering of the water table beyond the reach of existing wells. As a consequence, wells must be drilled deeper to reach the groundwater; in some places (e.g., California, Texas, and India) the water table has dropped hundreds of feet because of extensive well pumping. In the Punjab region of India, groundwater levels have dropped 10 meters since 1979, and the rate of depletion is accelerating. A lowered water table may, in turn, cause other problems such as groundwater-related subsidence and saltwater intrusion.

Wetlands contrast the arid landscape around Middle Spring, Fish
Springs National Wildlife Refuge, Utah

Subsidence

Subsidence occurs when too much water is pumped out from underground, deflating the space below the above-surface, and thus causing the ground to collapse. The result can look like craters on plots of land. This occurs because, in its natural equilibrium state, the hydraulic pressure of groundwater in the pore spaces of the aquifer and the aquitard supports some of the weight of the overlying sediments. When groundwater is removed from aquifers by excessive pumping, pore pressures in the aquifer drop and compression of the aquifer may occur. This compression may be partially recoverable if pressures rebound, but much of it is not. When the aquifer gets compressed, it may cause land subsidence, a drop in the ground surface. The city of New Orleans, Louisiana is actually below sea level today, and its subsidence is partly caused by removal of groundwater from the various aquifer/aquitard systems beneath it. In the first half of the 20th century, the city of San Jose, California dropped 13 feet from land subsidence caused by overpumping; this subsidence has been halted with improved groundwater management.

Contamination of Groundwater

Groundwater can become contaminated in many ways. If surface water that recharges an aquifer is polluted, the groundwater will also become contaminated. Contaminated groundwater can then affect the quality of surface water at discharge areas. Groundwater can also become contaminated when liquid hazardous substances soak down through the soil into groundwater.

Contaminants that can dissolve in groundwater will move along with the water, potentially to wells used for drinking water. If there is a continuous source of contamination entering moving groundwater, an area of contaminated groundwater, called a plume, can form. A combination of moving groundwater and a continuous source of contamination can, therefore, pollute very large volumes and areas of groundwater. Some plumes at Superfund sites are several miles long. More than 88 percent of current Superfund sites have some groundwater contamination.

Contamination of Groundwater by Liquids

Some hazardous substances dissolve very slowly in water. When these substances seep into groundwater faster than they can dissolve, some of the contaminants will stay in liquid form. If the liquid is less dense than water, it will float on top of the water table, like oil on water. Pollutants in this form are called light non-aqueous phase liquids (LNAPLs). If the liquid is more dense than water, the pollutants are called dense non-aqueous phase liquids (DNAPLs). DNAPLs sink to form pools at the bottom of an aquifer. These pools continue to contaminate the aquifer as they slowly dissolve and are carried away by moving groundwater. As DNAPLs flow downward through an aquifer, tiny globs of liquid become trapped in the spaces between soil particles. This form of groundwater contamination is called residual contamination.

Diagram 3
Contaminated Groundwater

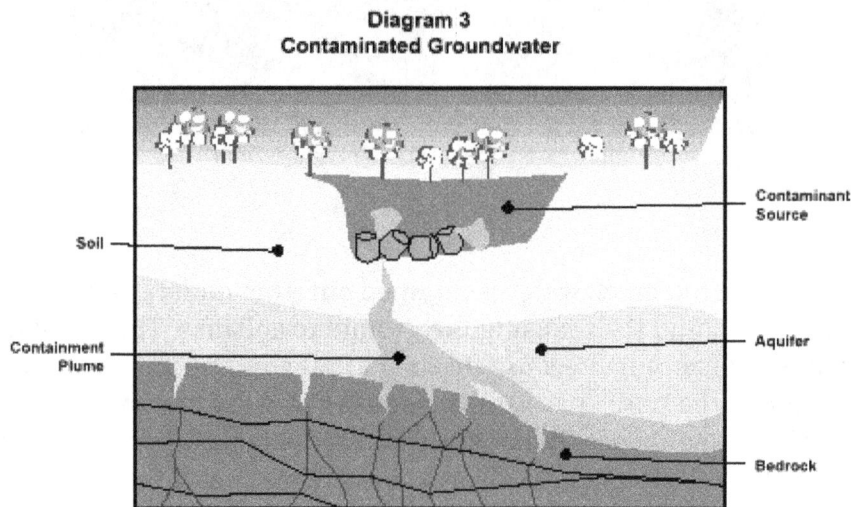

Factors That Affects Groundwater Contamination

Many processes can affect how contamination spreads and what happens to it in the groundwater, potentially making the contaminant more or less harmful, or toxic. Some of the most important processes affecting hazardous substances in groundwater are advection, sorption, and biological degradation.

- Advection occurs when contaminants move with the groundwater. This is the main form of contaminant migration in groundwater.

- Sorption occurs when contaminants attach themselves to soil particles. Sorption slows the movement of contaminants in groundwater, but also makes it harder to clean up contamination.

- Biological degradation happens when microorganisms, such as bacteria and fungi, use hazardous substances as a food and energy source. In the process, contaminants break down and hazardous substances often become less harmful.

Cleaning up Groundwater

Cleaning up contaminated groundwater often takes longer than expected because groundwater systems are complicated and the contaminants are invisible to the naked eye. This makes it more difficult to find contaminants and to design a treatment system that either destroys the contaminants in the ground or takes them to the surface for cleanup. Groundwater contamination is the reason for most of Superfund's long-term cleanup actions. Figure below illustrates groundwater treatment in action.

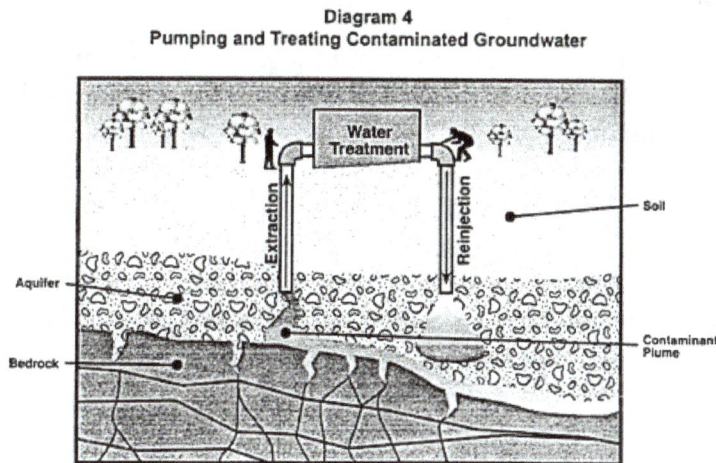

Figure: Pumping and Treating contaminated Groundwater.

Groundwater flow is the movement of water through interconnected voids in the phreatic zone. Countless measurements confirm that groundwater enjoys the latter fate groundwater indeed flows, and in some cases it moves great distances underground.

In the unsaturated zone the region between the ground surface and the water table water percolates straight down, like the water passing through a drip coffee maker, for this water moves only in response to the downward pull of gravity. But in the zone of saturation the region below the water table water flow is more complex, for in addition to the downward pull of gravity, water responds to differences in pressure. Pressure can cause groundwater to flow sideways, or even upward. Thus, to understand the nature of groundwater flow, we must first understand the origin of pressure in groundwater. For simplicity, we'll consider only the case of groundwater in an unconfined aquifer.

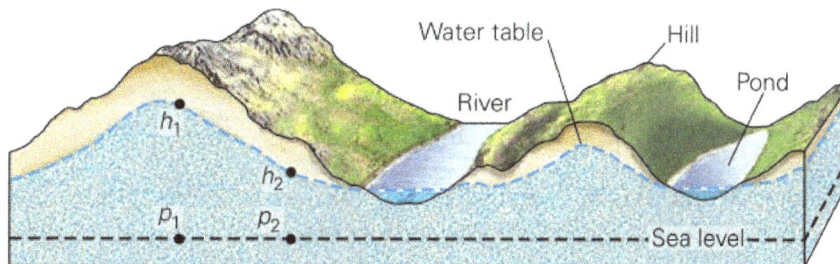

The shape of water table beneath hilly topography.

Pressure in groundwater at a specific point underground is caused by the weight of all the overlying water from that point up to the water table. (The weight of overlying rock does not contribute to the pressure exerted on groundwater, for the contact points between mineral grains bear the rock's weight.) Thus, a point at a greater depth below the water table feels more pressure than does a point at lesser depth. If the water table is horizontal, the pressure acting on an imaginary horizontal reference plane at a specified depth below the water table is the same everywhere. But if the water table is not horizontal, as shown in above, the pressure at points on a horizontal reference plane at depth changes with location. For example, the pressure acting at point p1, which lies below the hill in figure above, is greater than the pressure acting at point p2, which lies below the valley, even though both p1 and p2 are at the same elevation.

Both the elevation of a volume of groundwater and the pressure within the water provide energy that, if given the chance, will cause the water to flow. Physicists refer to such stored energy as potential energy. The potential energy available to drive the flow of a given volume of groundwater at a location is called the hydraulic head. To measure the hydraulic head at a point in an aquifer, hydrogeologists drill a vertical hole down to the point and then insert a pipe in the hole. The height above a reference elevation (for example, sea level) to which water rises in the pipe represents the hydraulic head water rises higher in the pipe where the head is higher. As a rule, groundwater flows from regions where it has higher hydraulic head to regions where it has lower hydraulic head. This statement generally implies that groundwater regionally flows from locations where the water table is higher to locations where the water table is lower.

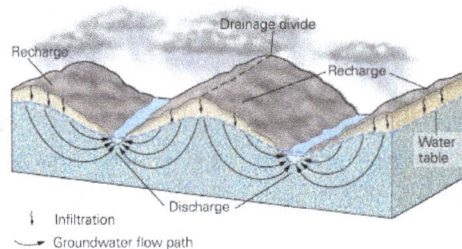

(a) Groundwater flows from recharge areas to discharge areas. Typically, the flow follows curving paths

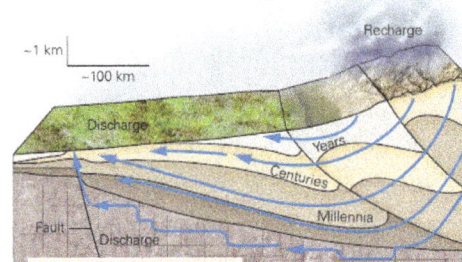

(b) The large hydraulic head resulting from uplift of a mountain belt may drive groundwater hundreds of kilometers, across regional sedimentary basins. Deeper flow paths take longer.

The flow of groundwater

Hydrogeologists have calculated how hydraulic head changes with location underground, by taking into account both the effect of gravity and the effect of pressure. These calculations reveal that groundwater flows along concave-up curved paths, as illustrated in cross section (figure above a, b). These curved paths eventually take groundwater from regions where the water table is high (under a hill) to regions where the water table is low (below a valley), but because of flow-path shape,

Some groundwater may flow deep down into the crust along the first part of its path and then may flow back up, toward the ground surface, along the final part of its path. The location where water enters the ground (where the flow direction has a downward trajectory) is called the recharge area, and the location where groundwater flows back up to the surface is called the discharge area.

Flowing water in an ocean current moves at up to 3 km per hour, and water in a steep river channel can reach speeds of up to 30 km per hour. In contrast, groundwater moves at less than a snail's pace, between 0.01 and 1.4 m per day (about 4 to 500 m per year). Groundwater moves much more slowly than surface water, for two reasons. First, groundwater moves by percolating through a complex, crooked network of tiny conduits, so it must travel a much greater distance than it would if it could follow a straight path. Second, friction between groundwater and conduit walls slows down the water flow.

Simplistically, the velocity of groundwater flow depends on the slope of the water table and the permeability of the material through which the groundwater is flowing. Thus, groundwater flows faster through high-permeability rocks than it does through low-permeability rocks and it flows faster in regions where the water table has a steep slope than it does in regions where the water table has a gentle slope. For example, groundwater flows relatively slowly (2 m per year) through a low-permeability aquifer under the Great Plains, but flows relatively quickly (30 m per year) through a high-permeability aquifer under a steep hill slope. In detail, hydrogeologists use Darcy's Law to determine flow rates at a location.

Darcy's Law for Groundwater Flow

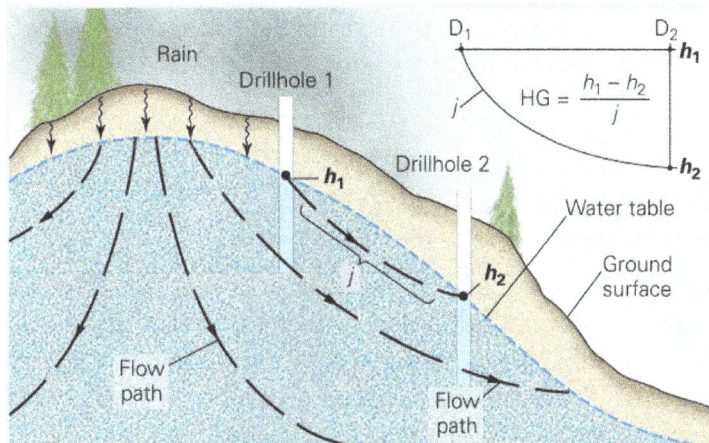

The level to which water rises in a drill hole is the hydraulic head (h).
The hydraulic gradient (HG) is the difference in head divided by the length of the flow path.

The rate at which groundwater flows at a given location depends on the permeability of the material containing the groundwater; groundwater flows faster in a more permeable material than it does in a less permeable material. The rate also depends on the hydraulic gradient, the change in hydraulic head per unit of distance between two locations, as measured along the flow path.

To calculate the hydraulic gradient, we divide the difference in hydraulic head between two points by the distance between the two points as measured along the flow path. This can be written as a formula:

hydraulic gradient = h1 - h2/j

where h1 - h2 is the difference in head (given in meters or feet, because head can be represented as an elevation) between two points along the water table, and j is the distance between the two points as measured along the flow path. A hydraulic gradient exists anywhere that the water table has a slope. Typically, the slope of the water table is so small that the path length is almost the same as the horizontal distance between two points. So, in general, the hydraulic gradient is roughly equivalent to the slope of the water table.

In 1856, a French engineer named Henry Darcy carried out a series of experiments designed to characterize factors that control the velocity at which groundwater flows between two locations (1 and 2), each of which has a different hydraulic head (h1 and h2). Darcy represented the velocity of flow by a quantity called the discharge (Q), meaning the volume of water passing through an imaginary vertical plane perpendicular to the groundwater's flow path in a given time. He found that the discharge depends on the the hydraulic head (h1- h2); the area (A) of the imaginary plane through which the groundwater is passing; and a number called the hydraulic conductivity (K). The hydraulic conductivity represents the ease with which a fluid can flow through a material. This, in turn, depends on many factors (such as the viscosity and density of the fluid), but mostly it reflects the permeability of the material. The relationship that Darcy discovered, now known as Darcy's law, can be written in the form of an equation as:

Q = KA(h1 - h2)/j

The equation states that if the hydraulic gradient increases, discharge increases, and that as conductivity increases, discharge increases. Put in simpler terms, the flow rate of groundwater increases as the permeability increases and as the slope of the water table gets steeper.

Groundwater Energy Balance

While flowing, the groundwater loses energy because of friction of flow, i.e. hydraulic energy is became heat. The groundwater energy balance may be the power balance of any groundwater body when it comes to inward hydraulic power associated with groundwater inflow into the body, power for this output, power alteration directly into warmth because of rubbing of circulation, as well as the ending adjust of energy standing as well as groundwater amount.

Theory

When multiplying the horizontal velocity of groundwater (dimension, for example, m^3/day per m^2 cross-sectional area) with the groundwater potential (dimension energy per m^3 water, or E/m^3) one obtains an energy flow (flux) in E/day per m^2 cross-sectional area.

Summation or integration of the energy flux in a vertical cross-section of unit width (say 1 m) from the lower flow boundary (the impermeable layer or base) up to the water table in an unconfined aquifer gives the energy flow Fe through the cross-section in E/day per m width of the aquifer.

While flowing, the groundwater loses energy due to friction of flow, i.e. hydraulic energy is converted into heat. At the same time, energy may be added with the recharge of water coming into the aquifer through the water table. Thus one can make an hydraulic energy balance of a block of soil between two nearby cross-sections. In steady state, i.e. without change in energy status and without accumulation or depletion of water stored in the soil body, the energy flow in the first section plus the energy added by groundwater recharge between the sections minus the energy flow in the second section must equal the energy loss due to friction of flow.

In mathematical terms this balance can be obtained by differentiating the cross-sectional integral of Fe in the direction of flow using the Leibniz rule, taking into account that the level of the water table may change in the direction of flow. The mathematics is simplified using the Dupuit–Forchheimer assumption.

The hydraulic friction losses can be described in analogy to *Joule's law* in electricity, where the friction losses are proportional to the square value of the current (flow) and the electrical resistance of the material through which the current occurs. In groundwater hydraulics (fluid dynamics, hydrodynamics) one often works with hydraulic conductivity (i.e. permeability of the soil for water), which is inversely proportional to the hydraulic resistance.

The resulting equation of the energy balance of groundwater flow can be used, for example, to calculate the shape of the water table between drains under specific aquifer conditions. For this a numerical solution can be used, taking small steps along the impermeable base. The drainage equation is to be solved by trial and error (iterations), because the hydraulic potential is taken with respect to a reference level taken as the level of the water table at the water divide midway between the drains. When calculating the shape of the water table, its level at the water divide is initially not known. Therefore, this level is to be assumed before the calculations on the shape of the water table can be started. According to the findings of the calculation procedure, the initial assumption is to be adjusted and the calculations are to be restarted until the level of the water table at the divide does not differ significantly from the assumed level.

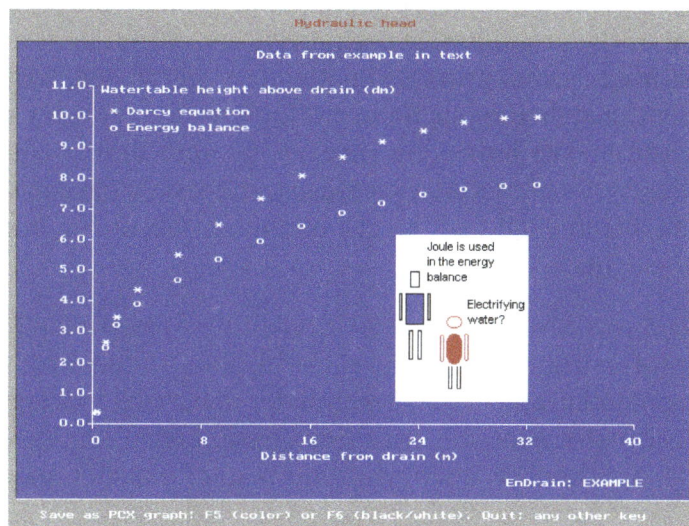

Shape of water table between drains

The trial and error procedure is cumbersome and therefore computer programs may be developed to aid in the calculations.

Application

The energy balance of groundwater flow can be applied to flow of groundwater to subsurface drains. The computer program *EnDrain* compares the outcome of the traditional drain spacing equation, based on Darcy's law together with the continuity equation (i.e. conservation of mass), with the solution obtained by the energy balance and it can be seen that drain spacings are wider in the latter case. This is owing to the introduction of the energy supplied by the incoming recharge.

Vadose Zone

This zone also includes the capillary fringe above the water table, the height of which will vary according to the grain size of the sediments. In coarse-grained mediums the fringe may be flat at the top and thin, whereas in finer grained material it will tend to be higher and may be very irregular along the upper surface. The vadose zone varies widely in thickness, from being absent to many hundreds of feet, depending upon several factors. These include the environment and the type of earth material present. Water within this interval, which is moving downward under the influence of gravity, is called vadose water, or gravitational water.

Vadose Zone Monitoring System

Vadose zone monitoring systems detect and or characterize changes in the zone between ground surface and the water table).

The first was developed to monitor contaminant changes after a site is remediated. The standard practice is to monitor groundwater wells for contaminants. This approach may be problematic at a site with a relatively deep groundwater table. By the time contaminants are detected in the groundwater, significant vadose zone contamination will have occurred. Therefore, monitoring the vadose zone instead of the groundwater permits earlier detection of a contaminant release. The Vadose Zone Monitoring System provides unattended, automated, "real-time" monitoring of soil vapor. An instrument, installed in single or multiple wells, provides gas sampling at up to 64 sampling ports. Measurements indicating changes in vapor movement suggest contaminant movement. This sampling system identifies, measures, and stores the concentration of up to five target gases. It shines an infrared light on the gas sample in a chamber, and sensors monitor the response of the gas to the light.

The second type of system was developed to detect contaminant migration from buried wastes. In this system, several types of instruments were developed to obtain *in situ* hydrologic characterization data, to verify drainage potential, and to obtain estimates of current recharge fluxes under a range of surface conditions. One instrument, the Advanced Tensiometer (AT), measures soil water pressures in the vadose zone. Another new instrument, the water fluxmeter, measures drainage flux.

Limitations and Concerns

In the first system, success depends upon gas vapor reaching the sampling locations. Barriers to vapor may give the false impression that a problem is not present. Also, the gas sampler can detect only five target gases. While vadose zone monitoring is important for soil contamination sites, risks associated with groundwater contamination may not be detected accurately.

With the second system, actual groundwater monitoring should be considered, even though hydrological monitoring of the vadose zone is important.

References

- Hydrogeology: yunus.hacettepe.edu.tr, Retrieved 11 June 2018

- Groundwater-Issues: newworldencyclopedia.org, Retrieved 19 April 2018

- Groundwater-flow: geologylearn.blogspot.com, Retrieved 30 June 2018

- Groundwater-energy-balance, engineering, science: assignmentpoint.com, Retrieved 26 July 2018

- Vadose-zone, science: britannica.com, Retrieved 15 May 2018

Evaporation and Transpiration

The transformation of water from a liquid phase to a gaseous phase during its movement from water bodies to the atmosphere is known as evaporation. The process of the release of water vapor from the soil and plants into the atmosphere is called transpiration. A detailed analysis of the fundamental concepts of evaporation and transpiration has been provided in this chapter, which includes topics such as pan evaporation, potential evaporation, transpiration and evapotranspiration.

Evaporation

Evaporation is also known as vaporization. Evaporation happens when a liquid substance becomes a gas. When water is heated, it evaporates. The molecules move and vibrate so quickly that they escape into the atmosphere as molecules of water vapor.

Evaporation is a very important part of the water cycle. Heat from the sun, or solar energy, powers the evaporation process. It soaks up moisture from soil in a garden, as well as the biggest oceans and lakes. The water level will decrease as it is exposed to the heat of the sun.

Water on the planet is constantly getting recycled in what's called the water cycle (or hydrological cycle), and evaporation is vital in that. Water from oceans, rivers, swamps, lakes, plants and even humans is converted to vapor in the water cycle, The sun provides solar energy that powers the evaporation process. The heat soaks up the moisture from soil in gardens, farms, oceans, and lakes. As a result, the water levels decrease due to exposure of heat from the sun, Though water levels in water bodies appear to decrease due to the sun's heat, the escaped molecules don't disappear. They stay in the atmosphere, and affect humidity and influence the moisture amounts in the air. Regions with high temperatures and large water bodies are humid due to water evaporating and remaining in the air as vapor. Evaporation also helps in cloud formation. Afterwards the clouds release the moisture as precipitation. In plants, transpiration is water evaporation from plants. In transpiration, water or minerals are carried from the roots, to the underside pores on the leaves in a plant. From these pores water evaporates into the atmosphere and that helps keep a plant cool, during hot weathers.

Utilization of Evaporation

For thousands of years, evaporation has been utilized in livelihoods. Table salt is obtained from saline sea water through evaporation, in evaporation ponds. The Dead Sea in the Middle East which has no river outlet relies on evaporation for water to leave this lake. Evaporation at the Dead Sea also aids in magnesium, potash and bromine minerals extraction according to the U.S. Geological Survey. Around 80 percent of all water evaporation comes from the oceans and the rest from inland water and vegetation.

Although the level of a lake, pool, or glass of water will decrease due to evaporation, the escaped water molecules dont disappear. They stay in the atmosphere, affecting humidity, or the amount of moisture in the air. Areas with high temperatures and large bodies of water, such as tropical islands and swamps, are usually very humid for this reason. Water is evaporating, but staying in the air as a vapor.

Once water evaporates, it also helps form clouds. The clouds then release the moisture as rain or snow. The liquid water falls to Earth, waiting to be evaporated. The cycle starts all over again.

Factors That Affect Vaporization

The main factors that have an effect on vaporization are as follows:

- Vaporization increases with an increase in the surface area.

- It increases with an increase in temperature.

- Vaporization increases with the decrease in humidity.

- It increases with an increase in wind speed.

Evaporation Increases with an Increase in the Surface Area

If the surface area is increased, then the amount is of liquid that is exposed to air is larger. More molecules can escape with a wider surface area. For e.g. we spread out clothes to dry. We do that because that speeds up the process of vaporization.

Evaporation Increases with an Increase in Temperature

The water molecules move rapidly when the water is heated. This makes the molecules escape faster. Higher temperatures lead to increase in vaporization as more molecules get kinetic energy to convert into vapor. For example, boiling water evaporates faster than fresh tap water.

Evaporation Increases With a Decrease in Humidity

Humidity means the amount of vapor present in the air. The air around can only hold a certain amount of vapor at a certain time and certain temperature. If the temperature increases and the wind speed and humidity stay constant, then the rate of evaporation will increase since warmer air can hold more water vapor than colder air.

Evaporation Increases With an Increase in the Speed of Wind

Particles of vapor move away when the speed of wind increases. This leads to a decrease in the amount of water vapor in the atmosphere. For example, we use hand dryers to dry our hands. Here the wind is expelled from the hand dryer which dries our hand.

Pan Evaporation

Pan evaporation is a weather measurement system that integrates several climatic conditions including rainfall, humidity, solar radiation, wind, temperature and drought dispersion. The system distinguishes the rates of evaporation based on the weather factor. The rate of evaporation is highest during hot, sunny, windy, and dry days and low during cloudy, calm, and humid weather. Although pan evaporation is no longer popular among researchers and scientist due to the emergence of better and more accurate technologies, pan evaporation is popular among farmers who seek to determine the amount of water required by their crops.

Evaporation Pans

An evaporation pan holds the water used during the process. The observer notes the quantity of water at certain weather conditions and notes the change in the quantity. Pans occur in different sizes and shapes, the most common being circular and square. "Class A" and the Sunken Colorado Pan are the most common in North America, but he Symon's Pan is popular in India, Europe, and South Africa. Advanced water pans are automated. They are fitted with water level sensors.

Class A Evaporation Pan

The U.S National Weather Service recommends the Class A evaporation pan. The pan contains cylinder that is 46.5 inches in diameter and 10 inches in depth. The pam is placed on a leveled wooden base and is enclosed by a chain-link fence to avoid interference by animals and insects. The rate of evaporation is determined daily by recording the depth of water. The initial quantity of the water is set at exactly two inches; at the end of the day, the water is then refilled. The amount of water it takes to fill the pan back to two inches is the rate of evaporation. The Class A Evaporation Pan is ineffective when the level of rainfall is beyond 30mm unless it is emptied several times in a 24 hour period. Past recordings using the pan have revealed that areas that experience heavy rainfall in excess of 30mm experience higher rates of evaporation on a daily basis than months where conditions suitable for evaporations prevail. When rainfall more than 55 mm is recorded, the pan is likely to overflow.

Sunken Colorado Pan

The sunken Colorado pan is square shaped. It is 3 ft. in width with a depth of 18 inches. The pan is made of galvanized iron. It is buried in the ground to a depth of 2 inches of its rim.

Global Warming and Pan Evaporation

It is theorized that global warming leads to increased rate of evaporation which in turn leads to accelerated hydrological cycle. With desertification spreading and weather patterns changing rapidly, it would take several years to determine to determine the change in the rate of evaporation on earth.

Decreasing Trend of Pan Evaporation

Over the last 50 or so years, pan evaporation has been carefully monitored. For decades, pan evaporation measurements were not analyzed critically for long term trends. But in the 1990s scientists reported that the rate of evaporation was falling. According to data, the downward trend had been observed all over the world except in a few places where it has increased.

It is currently theorized that, all other things being equal, as the global climate warms evaporation would increase proportionately and as a result, the hydrological cycle in its most general sense is bound to accelerate. The downward trend of pan evaporation has since also been linked to a phenomenon called global dimming. In 2005 Wild et al. and Pinker et al. found that the "dimming" trend had reversed since about 1990.

Other theories suggest that measurements have not taken the local environment into account. Since the local moisture level has increased in the local terrain, less water evaporates from the pan. This leads to false measurements and must be compensated for in the data analysis. Models accounting for additional local terrain moisture match global estimates.

Lake Evaporation vs. Pan Evaporation

Pan evaporation is used to estimate the evaporation from lakes. There is a correlation between lake evaporation and pan evaporation. Evaporation from a natural body of water is usually at a lower rate because the body of water does not have metal sides that get hot with the sun, and while light penetration in a pan is essentially uniform, light penetration in natural bodies of water will decrease as depth increases. Most textbooks suggest multiplying the pan evaporation by 0.75 to correct for this.

Potential Evaporation

The evaporation plus transpiration from a vegetated surface with unlimited water supply is known as potential evaporation and it constitutes the maximum possible rate due to the prevailing meteorological conditions. Thus PE is the maximum value of the actual evaporation (E_t): PE = E_t when water supply is unlimited.

Measurement of Potential Evaporation

There are many different ways of measuring evaporation. One of the most common methods is to use the irrigated lysimeter. Other ones are the use of an atmometer and the standardised US Class A pan.

Irrigated Lysimeter

Potential evaporation, the evaporation plus transpiration from a vegetated surface when water supply is unlimited, can be measured using irrigated lysimeters.

The installations used, shown in following figure, closely looks like a percolation gauge. The principal difference is in the operation of the apparatus, with the contained soil being kept at field capacity (the water content of the soil after the saturated soil has drained under gravity to equilibrum) by sprinkling a known quantity of water on the tank when rainfall is deficient.

Field capacity is assured by maintaining continuous percolation from the bottom of the tank. Thus the vegetation cover is allowed to transpire freely, and the total evaporation loss is dependent entirely on the ability of the air to absorb the water vapour. Then:

PE = Rainfall + Irrigation − Percolation

Difficulties in Operating

One of the disadvantages of these gauges is that the soil sample is disturbed, but with careful filling of the tank and after the establishment of the vegetation cover, the gauge gradually becomes representative of the surrounding ground.

In winter with snow cover and freezing temperatures, certain difficulties in operating the gauges are encountered, but discrepancies are not of great importance since evaporation losses are low and often negligible under such conditions.

Measured values of PE using these irrigated gauges can be exaggerated in very dry periods and hot climates. Surrounding parches ground heating and drying the air above tends to cause increased evaporation from the continuously watered and transpiring vegetation of the gauge.

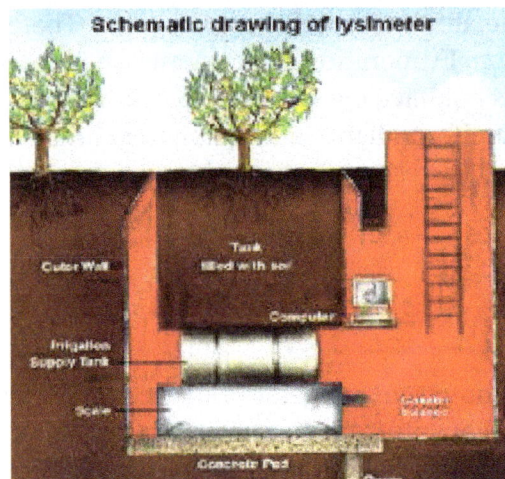

Schematic drawing of lysimeter

US Class a Pan

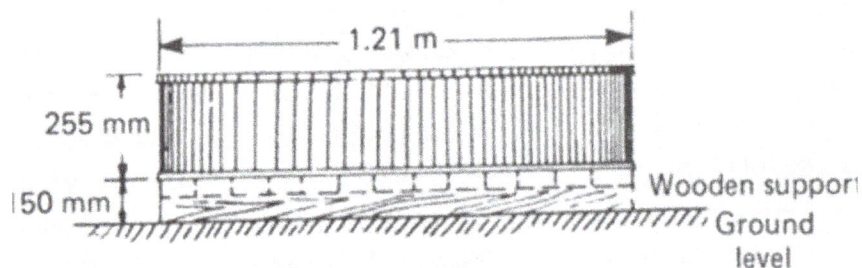

(b) US Class A pan

There are a lot of standardized pans for measuring evaporation and the US Class A pan is probably the most used. The pan is circular with a diameter of 1.21 m and depth of 255 mm which gives it a volume of about 0.3 m³. The basin is put on a 150 mm high wooden frame due to air circulation around the basin. The water level is kept about 50 mm below the rim, due to allowance of percolation and the need of water. The water level is measured every day, either you measure the difference between the present and the origin water level or if you have chosen to obtain the water level in the pan, you measure the amount of water you've put into the pan.

Due to that the sun hits the sides of the pan, the temperature gets higher which means that the evaporation gets higher than the actual evaporation. To correct this value you multiply your evaporation value from the pan with a coefficient, called pan coefficient and it's value depends on what climate region your test have been taken.

Another kind of pan is the UK British Standard tank which is a bit larger than the American pan and put on the same level as the ground. The principle is about the same as with the American pan. Though the general opinion in the UK is that measurement is unreliable and calculations are more common.

Atmometer

This is a device that can give direct measurement of evaporation. Atmometers basically consist of a wet, porous ceramic cup mounted on top of a cylindrical water reservoir. The ceramic cup is covered with a green fabric that simulates the canopy of a crop. The reservoir is filled with distilled water that evaporates out of the ceramic cup and is pulled through a suction tube that extends to the bottom of the reservoir. Underneath the fabric, the ceramic cup is covered by a special membrane that keeps rain water from seeping into the ceramic cup. A rigid wire extending from the top keeps birds from perching on top of the gauge.

Atmometer placed between irrigated fields.

Calculation of Potential Evaporation

A value of the actual evapotranspiration (E_t) over a catchment is more often obtained by first calculating the potential evapotranspiration (PE), assuming an unrestricted availability of water, and then modifying the answer by accounting for the actual soil moisture content.

There are several formulae for calculating potential evaporation (based on theoretical or empirical models), but the most commonly used are the following ones.

Penman Equation

This equation directly results from the basic formula which allowed to estimate evaporation from an open water surface. Then:

$$PE = \left(\frac{\Delta}{\Delta + \gamma}\right) * Q_{ET} + \left(\frac{\gamma}{\Delta + \gamma}\right) * E_{at} \quad \text{(mm/day)}$$

Where:

$Q_{ET} = Q_s*(1 - r) - Q_l$

$Q_l = 0.95*[8.64*10^7/(\rho*\lambda)]*\sigma*(273.16+Ta)^4*(0.53+0.065*(e_d-1.0)^{1/2})*(0.10+0.90*(n/N))$

$E_{at} = 0.3*(1+0.5*u_2)*(e_a-e_d)$

Δ (mb/C) is the the slope of the saturation vapour pressure curve with respect to temperature.

γ is the hygrometric constant (=0.65 mb/C).

Q_l is long wave radiation from the water body.

r is a coefficient relating to vegetation cover (r = 0.25 for a short grassed surface).

Ta is air temperature (C).

n/N is the ratio of actual/possible sunshine hours of bright sunshine.

ρ is the density of water (kg/m³).

λ is the latent heat of evaporization of water (J/kg).

σ is Stefan Bolzman's constant (= $5.7*10^{-8}$ W/(m²*grad⁴)).

u_2 is wind velocity (m/s).

e_a is the saturation vapour pressure for the measured air temperature (mb).

e_d is the actual vapour pressure of the air (mb).

OBS: Q_{ET}, Q_s, Q_l, E_{at} are all expressed in mm/day.

Thornthwaite's Formula

This formula is based mainly on temperature with an adjustment being made for the number of daylight hours. An estimate of the potential evapotransiration, calculated on a monthly basis, is given by:

$$PE_m \quad 16N_m \left(\frac{10\,T}{}\right) \quad \text{mm}$$

where m is the months 1, 2, 3...12, N_m is the monthly adjustment factor related to hours of daylight, T_m is the monthly mean temperature (C), I is the heat index for the year, given by: $I = \Sigma i_m = \Sigma \left(\dfrac{\bar{T}_m}{5}\right)^{1.5}$ for m = 1...12

and: $a = 6.7*10^{-7}*I^3 - 7.7*10^{-5}*I^2 + 1.8*10^{-2}*I + 0.49$

Given the monthly mean temperatures from the measurements at a climatologgical station, an estimate of the potential evaporation for each month of the year can be calculated. This method has been used widely throughout the world, but strictly is not valid for climates other than those similar to that area where it was developed (the eastern USA).

Compared to the Penman formula, Thornthwait values tend to exaggerate the potential evaporation. This is particularly marked in the summer months with the high temperatures having a dominant effect in the Tornthwaite computation, whereas the Penman estimate takes into consideration other meteorological factors.

Turc's Formula

Toward the needs of the agronomists for irrigation schemes, Turc extended his empirical method for calculating the annual actual evapotranspiration to produce a formula for potential evaporation over a shorter period of time.

The Turc short term formula for potential evaporation over 10 days is:

$$PE = \dfrac{P+a+70}{\left[1+\left(\dfrac{P+a}{L}+\dfrac{70}{2L}\right)^2\right]^{0.5}} \quad mm$$

where P is the precipitation in a 10 day period (mm), a is the estimated evaporation in the 10-day period from the bare soil when there has been no precipitation (1mm$\leq a \leq$ 10mm), and L is the "evaporation capacity" of the air given by:

$$L = \dfrac{(T+2)Q_s^{0.5}}{16}$$

with T the mean air temperature over the 10 days (C) and Q_s the mean short wave radiation (cal cm^{-2} day^{-1}).

Blaney-Criddle Formula

This formula, based on another empirical model, requires only mean daily temperatures T (C) over each month. Then:

$$PE = p.(0.46.T+8) \quad mm/day$$

where p is the mean daily percentage (for the month) of total annual daytime hours.

Actual Evaporation

Actual evaporation is the amount of water which is evaporated a normal day which means that if for instance the soil runs out of water, the actual evaporation is the amount of water which has been evaporated, and not the amount of water which could have been evaporated if the soil had had an infinite amount of water to evaporate.

To estimate a value on the actual evaporation for an area you have a couple of different choices. You could either put up a water balance for the area, measure the water flow or you could put up an equation of the potential evaporation times a function depending on the amount of water available.

Evaporation From Soils

The water which evaporates from a soil is stored in the small pores between the particles in the ground. When a rainfall hits the ground the pores are filled up with water, most of it infiltrates to the ground water or is led to a water course. From the point of field capacity, i.e. the amount of water which the soil manages to hold after free drainage, to the moment of wilking point, the water is accessible for evaporation (if the water content gets larger than the field capacity the process is called evaporation from free water surface). During this time we have had a pressure drop from corresponding -10 cm to -1000 m.

Transpiration

Water is not just evaporating from open water or from the soil, but also from vegetation. This is called transpiration, or together with the evaporation, evapotranspiration. Transpiration takes place through leaves pores, which are called stomata. The amount of vapour released from the pore depends of the temperature of the leaf, light and the amount of water in the leaf. All together these factors are called Stomata-resistance, r_s. The value r_s is raised at dry weather or if the content of water in the vegetation is low. We will write more about this factor further down when we write about the Penman-Monteith formula.

Calculation of Actual Evaporation

There are a few formulas which are more common than others. The origin of the equation is the Penman equation and later Monteith developed the formula even more.

Penman-Monteith

The most known formula for evapotranspiration is the Penman-Monteith formula,

$$E_T = \frac{\Delta R_n + (e_a - e_d) * \dfrac{\rho^* c_p}{r_a}}{\lambda (\Delta + \gamma^* (1 + \dfrac{r_s}{r_a}))}$$

where R_n = net radiation (W/m²)

ρ = density of air

c_p = specific heat of air

r_s = net resistance to diffusion through the surfaces of the leaves and soil (s/m)

r_a = net resistance to diffusion through the air from surfaces to height of   measuring instruments (s/m).

γ = hygrometric constant

$\Delta = de/dT$

e_a = saturated vapour pressure at air temperature

e_d = mean vapour pressure

The method is of quite good accuracy and is usually used for calculations of evapotranspiration from farmlands. The good accuracy is due to all the parameters of the equation but still it isn't perfect. For instance, the r_s value is a constant depending on what kind of vegetation the area holds. If the equation is used over a large area with different kind of vegetation you have to estimate a value for r_s. The estimation gets even more non accurate if the area contains spots without vegetation.

Methods for Measuring Actual Evaporation

Measuring actual evaporation is probably not as common as measuring potential. The most common method is the percolation gauge.

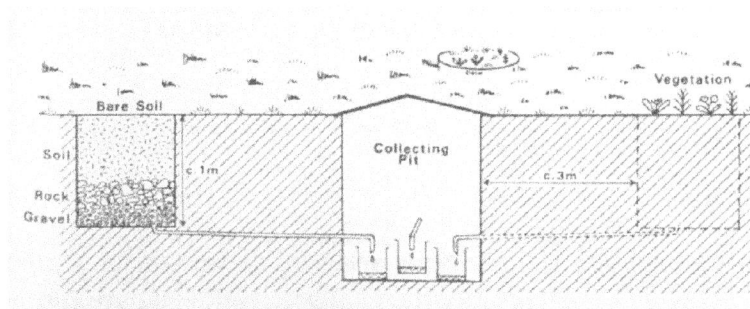

Picture of percolation gauge

The percolation gauge is actually regarded as a research tool and not a standard instrument for measuring evaporation and transpiration. There are many different designs of the gauge but the one on the picture is recommended. On the left side can we see a 1 meter deep hole filled up with soil, rock and gravel and a pipe from the bottom to the collection pit. The top of the hole should be indistinguishable from the surrounding vegetation.

When you measure evaporation with a percolation gauge, you take no consideration to changes in soil water storage. That means that the measurements should be made over a time period when the gauge is saturated.

Another method to measure evapotranspiration is with a lysimeter which takes consideration to how much water is stored in the soil. The lysimeter weigths the soil and gives a value on how much water is stored. This method is more complex, expensive and harder to maintain than percolation gauges.

Transpiration

Transpiration is the transfer of water into the air via leaf pores or *stomata*. Interestingly, the same three requirements for evaporation apply to transpiration. Approximately 600 cal/gm of energy is required to transpire water from leaves. The transfer of water into the air removes heat from the plant and so transpiration, like evaporation, is a cooling process. Thus transpiration is an important means of transporting heat between the surface and air above.

Water for transpiration comes from that which is stored in the soil and then extracted by plant roots. The amount of water that is held in the soil moisture zone is dependent on the texture and structure of the soil. Coarse textured soil dominated by sand-size particles holds less moisture than a finer textured soil.

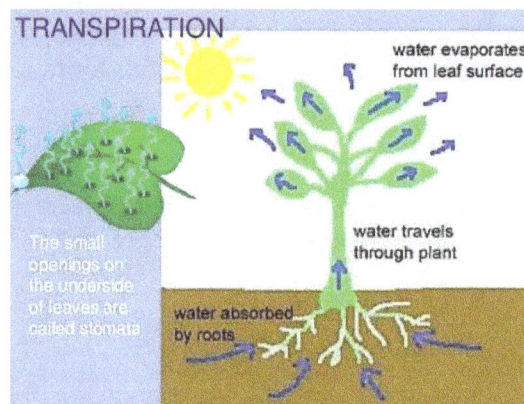

Process of Transpiration

Before the process of transpiration, there are a series of processes for a plant to undergo. After completing all these processes, transpiration takes place. Following are the steps which will help you understand the complete process.

- Plant takes water, dissolved essential plant nutrients and minerals from the soil with the help of the roots through the process of osmosis.

- Due to the lower water pressure in the leaves and upper part of the plants, the water travels from the roots to the upper parts through xylem.

- The water and the other minerals get mixed with the CO_2 and chlorophyll in the leaves and prepare food with the help of sunlight.

- Here, the process of Transpiration starts. When the water reaches the leaves, it is brought to the surface of the leaves with the help of stomata. Stomata help in the exchange of gases, that is, they take in CO_2 and give out O_2 in the atmosphere.

Role of Stomata

Stomata plays the lead role in conducting the process of transpiration. Stomata has two guard cells which are responsible for their opening and closing. The rate of transpiration is directly proportional to the opening and number of stomata. In the daytime, the stomata is open. As the sun is not present at night, the cells remain close at that part of the time. The stomata release water in the atmosphere, which is then broken down into oxygen and hydrogen. In return, the atmosphere gives carbon-dioxide to the plant to complete its process of photosynthesis.

The number of stomata may vary in different plants. In xerophytes the number of stomata will be less as compared to the other plants. This reduces the water loss and helps the plant to survive in adverse conditions. The plant may also close its leaves if there is excess sunlight, to save the water from transpiration.

Atmospheric Factors Affecting Transpiration

The amount of water that plants transpire varies greatly geographically and over time. There are a number of factors that determine transpiration rates:

- Temperature: Transpiration rates go up as the temperature goes up, especially during the growing season, when the air is warmer due to stronger sunlight and warmer air masses. Higher temperatures cause the plant cells which control the openings (stoma) where water is released to the atmosphere to open, whereas colder temperatures cause the openings to close.

- Relative humidity: As the relative humidity of the air surrounding the plant rises the transpiration rate falls. It is easier for water to evaporate into dryer air than into more saturated air.

- Wind and air movement: Increased movement of the air around a plant will result in a higher transpiration rate. Wind will move the air around, with the result that the more saturated air close to the leaf is replaced by drier air.

- Soil-moisture availability: When moisture is lacking, plants can begin to senesce (premature ageing, which can result in leaf loss) and transpire less water.

- Type of plant: Plants transpire water at different rates. Some plants which grow in arid regions, such as cacti and succulents, conserve precious water by transpiring less water than other plants.

Transpiration and Groundwater

In many places, the top layer of the soil where plant roots are located is above the water table and thus is often wet to some extent, but is not totally saturated, as is soil below the water table. The soil above the water table gets wet when it rains as water infiltrates into it from the surface, but, it will dry out without additional precipitation. Since the water table is usually below the depth of the plant roots, the plants are dependent on water supplied by precipitation. As this diagram shows, in places where the water table is near the land surface, such as next to lakes and oceans, plant roots can penetrate into the saturated zone below the water table, allowing the plants to transpire water directly from the groundwater system. Here, transpiration of groundwater commonly results in a drawdown of the water table much like the effect of a pumped well (cone of depression—the dotted line surrounding the plant roots in the diagram).

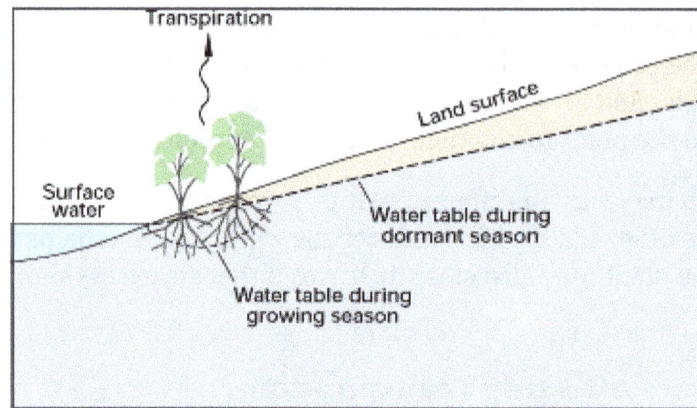

Evapotranspiration

Evapotranspiration is a combination of the words evaporation and transpiration. It refers to the process of water (moisture) moving from the soil and the plant, and entering the earth's atmosphere.

Evapotranspiration is really two processes: the evaporation of water from the soil, and the transpiration of moisture from the plant's surfaces. With this process, clouds and rain eventually occur.

Accurately studying and determining evapotranspiration rates in a greenhouse or growroom is a key component in determining irrigation requirements.

Factors such as the sunlight, relative humidity, air temperature, and wind speed all affect evapotranspiration. If sunlight, wind, and temperature levels increase but the humidity level drops, then the rate of evapotranspiration drops.

When the soil is dry, the rate of evapotranspiration lowers. Evapotranspiration is greatest if the surface of the plants and the soil are kept moist via overhead irrigation methods.

All plant life releases water vapor. Plants draw water and nutrients up from the soil and release the excess through their foliage into the atmosphere. This puts valuable moisture into the air and atmospheric conditions. The process is responsible for 15 per cent of all atmospheric moisture.

Without evapotranspiration, clouds could not form and rain would not fall.

Importance of Evapotranspiration

Water continuously moves between the oceans, sky and land. This ongoing circulation is fundamental to the availability of water on the planet and therefore to life on earth. ET is a key process within this cycle, and is responsible for 15% of the atmosphere's water vapor. Without it clouds couldn't form and rain wouldn't fall.

Estimating Evapotranspiration

Evapotranspiration can be measured or estimated using several methods.

Indirect Methods

Pan evaporation data can be used to estimate lake evaporation, but transpiration and evaporation of intercepted rain on vegetation are unknown. There are three general approaches to estimate evapotranspiration indirectly.

Catchment Water Balance

Evapotranspiration may be estimated by creating an equation of the water balance of a drainage basin. The equation balances the change in water stored within the basin (S) with inputs and outgoes:

$$\Delta S = P - ET - Q - D$$

The input is precipitation (P) and the outgoes are evapotranspiration (which is to be estimated), streamflow (Q), and groundwater recharge (D). If the change in storage, precipitation, streamflow, and groundwater recharge are all estimated, the missing flux, ET, can be estimated by rearranging the above equation as follows:

$$ET = P - \Delta S - Q - D$$

Energy Balance

A third methodology to estimate the actual evapotranspiration is the use of the energy balance.

$$\lambda E = R_n - G - H$$

where λE is the energy needed to change the phase of water from liquid to gas, R_n is the net radiation, G is the soil heat flux and H is the sensible heat flux. Using instruments like a scintillometer, soil heat flux plates or radiation meters, the components of the energy balance can be calculated and the energy available for actual evapotranspiration can be solved.

The SEBAL and METRIC algorithms solve the energy balance at the earth›s surface using satellite imagery. This allows for both actual and potential evapotranspiration to be calculated on a pixel-by-pixel basis. Evapotranspiration is a key indicator for water management and irrigation performance. SEBAL and METRIC can map these key indicators in time and space, for days, weeks or years.

Experimental Methods for Measuring Evapotranspiration

One method for measuring evapotranspiration is with a weighing lysimeter. The weight of a soil column is measured continuously and the change in storage of water in the soil is modeled by the change in weight. The change in weight is converted to units of length using the surface area of the weighing lysimeter and the unit weight of water. Evapotranspiration is computed as the change in weight plus rainfall minus percolation.

Remote Sensing

In recent decades, estimating evapotranspiration has been improved by advances in remote sensing, particularly in agricultural studies. However, quantifying evapotranspiration from mixed vegetation environs, particularly urban parklands, is still challenging because of the heterogeneity of plant species, canopy covers and microclimates and because the methodology is costly. Different remote sensing-based approaches for estimating evapotranspiration have various advantages and disadvantages.

Eddy Covariance

The most direct method of measuring evapotranspiration is with the eddy covariance technique in which fast fluctuations of vertical wind speed are correlated with fast fluctuations in atmospheric water vapor density. This directly estimates the transfer of water vapor (evapotranspiration) from the land (or canopy) surface to the atmosphere.

Hydrometeorological Equations

The most general and widely used equation for calculating reference ET is the Penman equation. The Penman-Monteith variation is recommended by the Food and Agriculture Organization and the American Society of Civil Engineers. The simpler Blaney-Criddle equation was popular in the Western United States for many years but it is not as accurate in regions with higher humidities. Other solutions used includes Makkink, which is simple but must be calibrated to a specific location, and Hargreaves. To convert the reference evapotranspiration to actual crop evapotranspiration, a crop coefficient and a stress coefficient must be used. Crop coefficients as used in many hydrological models usually change along the year to accommodate to the fact that crops are seasonal and, in general, plants behave differently along the seasons: perennial plants mature over multiple seasons, and stress responses can significantly depend upon many aspects of plant condition.

Urban Landscape Plants

Methods for measuring evapotranspiration can be adapted to an urban setting to estimate the water requirements of urban landscape vegetation.

Evapotranspiration of Urban Vegetation

The water requirement of urban landscapes, particularly urban parklands, is of growing concern. The estimation of evapotranspiration (ET) and subsequently plant water requirements in urban vegetation needs to consider the heterogeneity of plants, soils, water, and climate characteristics. In a research in South Australia, two practical ET estimation approaches are compared to a detailed Soil Water Balance (SWB) analysis over a one-year period. One approach is the Water Use Classification of Landscape Plants (WUCOLS) method, which is based on expert opinion on the water needs of different classes of landscape plants. The other is a remote sensing approach based on the Enhanced Vegetation Index (EVI) from Moderate Resolution Imaging Spectroradiometer (MODIS) sensors on the Terra satellite. Both methods require knowledge of reference ET calculated from meteorological data. More information is available in:

Comparing Three Approaches of Evapotranspiration Estimation in Mixed Urban Vegetation: Field-Based, Remote Sensing-Based and Observational-Based Methods *

Water requirements of urban landscape plants: a comparison of three factor-based approaches*

High Spatial Resolution WorldView-2 Imagery for Mapping NDVI and Its Relationship to Temporal Urban Landscape Evapotranspiration Factors*

Potential Evapotranspiration

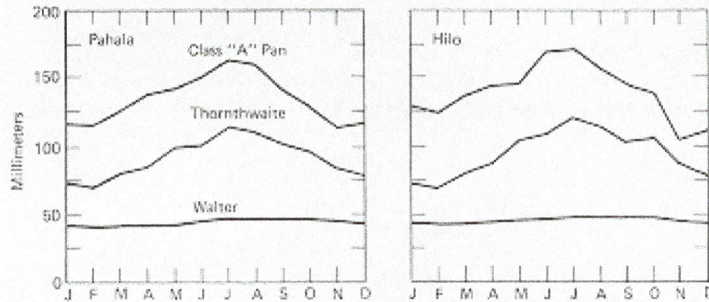

Monthly estimated potential evapotranspiration and measured pan
evaporation for two locations in Hawaii, Hilo and Pahala.

Potential evapotranspiration (PET) is the amount of water that would be evaporated and transpired by a specific crop or ecosystem if there were sufficient water available. This demand incorporates the energy available for evaporation and the ability of the lower atmosphere to transport evaporated moisture away from the land surface. Potential evapotranspiration is higher in the summer, on less cloudy days, and closer to the equator, because of the higher levels of solar radiation that provides the energy for evaporation. Potential evapotranspiration is also higher on windy days because the evaporated moisture can be quickly moved from the ground or plant surface, allowing more evaporation to fill its place.

Potential evapotranspiration is expressed in terms of a depth of water, and can be graphed during the year.

Potential evapotranspiration is usually measured indirectly, from other climatic factors, but also depends on the surface type, such as free water (for lakes and oceans), the soil type for bare soil, and the vegetation. Often a value for the potential evapotranspiration is calculated at a nearby climate station on a reference surface, conventionally short grass. This value is called the reference evapotranspiration, and can be converted to a potential evapotranspiration by multiplying with a surface coefficient. In agriculture, this is called a crop coefficient. The difference between potential evapotranspiration and precipitation is used in irrigation scheduling.

Average annual potential evapotranspiration is often compared to average annual precipitation, P. The ratio of the two, P/PET, is the aridity index.

Penman Equation

The Penman method (1948), applied to open water, can be briefly described by the energy balance at the earth's surface, which equates all incoming and outgoing energy fluxes. It reads

$$R_n = H + \lambda E + G$$

where

R_n = energy flux density of net incoming radiation (W/m²)

H   = flux density of sensible heat into the air (W/m²)

λE = flux density of latent heat into the air (W/m²)

G   = heat flux density into the water body (W/m²)

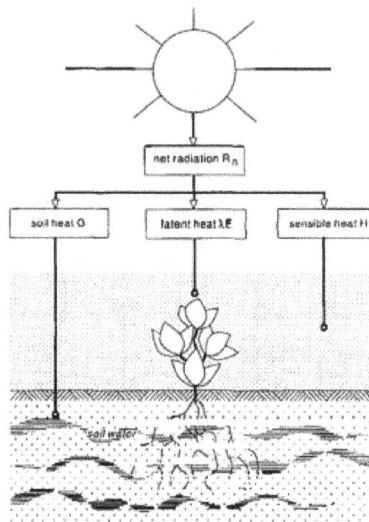

Figure: Illustration of the variables involved in the energy balance at the soil surface

The coefficient λ in is the latent heat of vaporization of water, and E is the vapour flux density in kg/m²s. To convert the above λE in W/m² into an equivalent evapo(transpi)ration in units of mm/d, λE should be multiplied by a factor 0.0353. This factor equals the number of seconds in a day (86 400), divided by the value of λ (2.45 x 10⁶ J/kg at 20°C), whereby a density of water of 1000 kg/m³ is assumed.

Supposing that R_n and G can be measured, one can calculate E if the ratio H/λE (which is called the Bowen ratio) is known. This ratio can be derived from the transport equations of heat and water vapor in air.

The situation depicted in figure above and described by equation ($R_n = H + \lambda E + G$) shows that radiation energy, R_n - G, is transformed into sensible heat, H, and water vapour, λE, which are transported to the air in accordance with:

$$H = c_1 \left(\frac{T_s - T_a}{r_a} \right)$$

$$\lambda E = c_2 \left(\frac{e_s - e_d}{r_a} \right)$$

where

c_1, c_2 = constants

T_s = temperature at the evaporating surface (°C)

T_a = air temperature at a certain height above the surface (°C)

e_s = saturated vapor pressure at the evaporating surface (kPa)

e_d = prevailing vapor pressure at the same height as Ta (kPa)

r_a = aerodynamic diffusion resistance, assumed to be the same for heat and water vapour   (s/m)

The slope Δ in Figure below can be determined at temperature Ta, provided that $(T_s - T_a)$ is small.

$$\frac{H}{\lambda E} = \frac{c_1(T_s - T_a)}{c_2(e_s - e_d)}$$

where

$$c_1/c_2 = \gamma = \text{psychrometric constant (kPa/°C)}$$

The problem is that generally the surface temperature, T_s, is unknown. Penman therefore introduced the additional equation

$$e_s - e_a = \Delta (T_s - T_a)$$

where the proportionally constant Δ (kPa/°C) is the first derivative of the function $e_s(T)$, known as the saturated vapour pressure curve. Note that e, in equation above is the saturated vapour pressure at temperature Ta. Re-arranging gives

$$\Delta = \frac{e_s - e_a}{T_s - T_a} \approx \frac{de_a}{dT_a}$$

The slope Δ in figure below can be determined at temperature Ta, provided that $(T_s - T_a)$ is small.

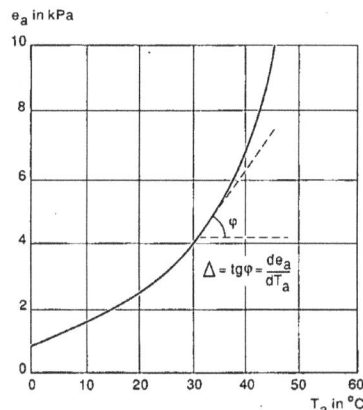

Figure: Saturated water vapour pressure e_a as a function of air temperature T_a

From Equation ($e_s - e_a = \Delta (T_s - T_a)$), it follows that $T_s - T_a = (e_s - e_a)/\Delta$. Substitution into

Equation ($\dfrac{H}{\lambda E} = \dfrac{c_1(T_s - T_a)}{c_2(e_s - e_d)}$) yields

$$\frac{H}{\lambda E} = \frac{\gamma}{\Delta} \frac{e_s - e_d}{e_s - e_a}$$

If ($e_s - e_a$) is replaced by ($e_s - e_d - e_a + e_d$), Equation ($\dfrac{H}{\lambda E} = \dfrac{\gamma}{\Delta} \dfrac{e_s - e_d}{e_s - e_a}$) can be written as

$$\frac{H}{\lambda E} = \frac{\gamma}{\Delta}\left[1 - \frac{e_a - e_d}{e_s - e_d}\right]$$

Under isothermal conditions (i.e. if no heat is added to or removed from the system), we can assume that $T_s \approx T_a$. This implies that $e_s \approx e_a$. If we then introduce this assumption, the isothermal evaporation, λE_a, reads as

$$\lambda E_a = c_2 \frac{e_a - e_d}{r_a}$$

Dividing Equation ($\lambda E_a = c_2 \dfrac{e_a - e_d}{r_a}$) by Equation ($\lambda E = c_2(\dfrac{e_s - e_d}{r_a})$) yields

$$\frac{E_a}{E} = \frac{e_a - e_d}{e_s - e_d}$$

The ratio on the right also appeared in equation ($\dfrac{H}{\lambda E} = \dfrac{\gamma}{\Delta}\left[1 - \dfrac{e_a - e_d}{e_s - e_d}\right]$), which can now be written as

$$\frac{H}{\lambda E} = \frac{\gamma}{\Delta}\left(1 - \frac{E_a}{E}\right)$$

From Equation ($R_n = H + \lambda E + G$), it follows that $H = R_n - \lambda E - G$. After some rearrangement, and writing E_o (subscript $\circ$ denoting open water) for E, substitution into equation ($\dfrac{H}{\lambda E} = \dfrac{\gamma}{\Delta}\left(1 - \dfrac{E_a}{E}\right)$) yields the formula of Penman (1948)

$$E_o = \frac{\Delta(R_n - G)/\lambda + \gamma E_a}{\Delta + \gamma}$$

where

E_o = open water evaporation rate (kg/m² s)

Δ = proportionality constant de_a/dT_a (kPa/°C)

R_n = net radiation (W/m²)

λ = latent heat of vaporization (J/kg)

γ = psychrometric constant (kPa/°C)

E_a = isothermal evaporation rate (kg/m² s)

The term $\dfrac{\Delta}{\Delta + \gamma}(R_n - G)/\lambda$ is the evaporation equivalent of the net flux density of radiant energy to the surface, also called the radiation term. The term $\dfrac{\Delta}{\Delta + \gamma}E_a$ is the corresponding aerodynamic term. equation ($\dfrac{\Delta(R_n - G)/\lambda + \gamma E_a}{\Delta + \gamma}$) clearly shows the combination of the two processes in one formula.

For open water, the heat flux density into the water, G, is often ignored, especially for longer periods. Also note that the resulting E_o in kg/m² s should be multiplied by 86 400 to give the equivalent evaporation rate E_o in mm/d.

E_o has been used as a kind of reference evaporation for some time, but the practical value of estimating E_o with the original Penman formula (Equation ($E_o = \dfrac{\Delta(R_n - G)/\lambda + \gamma E_a}{\Delta + \gamma}$)) is generally limited to large water bodies such as lakes, and, possibly, flooded rice fields in the very early stages of cultivation.

The modification to the Penman method introduced by Doorenbos and Pruitt started from the assumption that evapotranspiration from grass also largely occurs in response to climatic conditions. And short grass being the common surroundings for agrometeorological observations,they suggested that the evapotranspiration from 8 - 15 cm tall grass, not short of water, be used as a reference, instead of open water. The main changes in Penman's formula to compute this reference evapotranspiration, ET, (g for grass), concerned the short-wave reflection coefficient (approximately 0.05 for water and 0.25 for grass), a more sensitive wind function in the aerodynamic term, and an adjustment factor to take into account local climatic conditions deviating from an assumed standard.The adjustment was mainly necessary for deviating combinations of radiation, relative humidity, and day/night wind ratios; relevant values can be obtained from a table in Doorenbos and Pruitt .

If the heat flux, G, is set equal to zero for daily periods, the FAO Modified Penman equation can be written as

$$ET_g = c\left[\frac{\Delta}{\Delta + \gamma}R'_n + \frac{\gamma}{\Delta + \gamma}2.7f(u)(e_a - e_d)\right]$$

where

ET_g = reference evapotranspiration rate (mm/d)

C   = adjustment factor (-)

Rn'   = equivalent net radiation (mm/d)

f(u)   = wind function; $f(u) = 1 + 0.864\,u_2$

u_2 = wind speed (m/s)

$e_a - e_d$ = vapour pressure deficit (kPa)

Δ, γ = as defined earlier

Potential evapotranspiration from cropped surfaces was subsequently found from

appropriate crop coefficients, for the determination of which Doorenbos and Pruitt also provided a procedure.

Penman–Monteith Equation

Various derivations of the Penman equation included a bulk surface resistance term. The resulting equation is now called the Penman-Monteith equation, which may be expressed for daily values as

$$\lambda ET_o = \frac{\Delta(R_n - G) + \dfrac{86,400\,\rho_a C_P (e_s^{\,o} - e_a)}{r_{av}}}{\Delta + \gamma\left(1 + \dfrac{r_s}{r_{av}}\right)}$$

where ρa is air density in kg m^{-3}, Cp is specific heat of dry air [~1.013×10^{-3} MJ kg-1 ^{o}C^{-1}], $e_s^{\,o}$ is mean saturated vapor pressure in kPa computed as the mean e^{o} at the daily minimum and maximum air temperature in ^{o}C, r_{av} is the bulk surface aerodynamic resistance for water vapor in s m^{-1}, e_a is the mean daily ambient vapor pressure in kPa, and r_s is the canopy surface resistance in s m^{-1}. The Penman-Monteith equation represents the evaporating surface as a single "big leaf" with two parameters – one of which is determined by the atmospheric physics (r_{av}) influenced only slightly by the crop canopy architecture while the other one (r_s) depends on the biological behavior of the crop canopy surface and is related to both crop specific parameters (light attenuation, leaf stomatal resistances, etc.) and environmental parameters (irradiance, vapor pressure deficit, etc.). The water vapor aerodynamic resistance can be estimated following as,

$$r_{av} = \frac{\ln\left[\dfrac{(z_w - d)}{z_{om}}\right]\ln\left[\dfrac{(z_r - d)}{z_{ov}}\right]}{k^2 U_z}$$

where z_w is the wind speed measurement height in m, z_{om} is the momentum roughness length in m, z_r is the relative humidity measurement height in m, and z_{ov} is the vapor roughness length in m. The crop canopy aerodynamic parameters are estimated as follows,

$$d \quad (2/3)h$$
$$z_{om} \quad 0.123\ h_c$$
$$z_{av} \quad 0.1\ z_{om}$$

Equation mentioned above the last one is referenced here as the ASCE Penman-Monteith equation with all parameters computed as outlined by.

FAO-56 Penman-Monteith Equation

simplified eqn. ($\lambda ET_o = \dfrac{\Delta(R_n - G) + \dfrac{86,400\rho_a C_P (e_s{}^\circ e_a)}{r_{av}}}{\Delta + \gamma\left(1 + \dfrac{r_s}{r_{av}}\right)}$) by utilizing some assumed

constant parameters for a clipped grass reference crop that is 0.12-m tall in an extensive report for the Food and Agriculture Organization of the United Nations (FAO-56 Paper). They assumed the definition drafted by an FAO Expert Consultation Panel for the reference crop as "a hypothetical reference crop with an assumed crop height of 0.12 m, a fixed surface resistance of 70 s m^{-1} and an albedo of 0.23." By further assuming a constant for λ and simplifying the air density term (ρa), they derived the FAO-56 Penman-Monteith equation using the fixed bulk surface resistance (70 s m^{-1}) and the vapor aerodynamic resistance simplified to an inverse function of wind speed ($r_{av} = 208 / U_z$), as

$$ET_o = \frac{0.408\Delta(R_n - G) + \gamma \dfrac{900}{T + 273} U_2(e^\circ - e_a)}{\Delta + \gamma\left(1 + 0.34 U_2\right)}$$

where ET_o is the hypothetical reference crop evapotranspiration rate in mm d^{-1}, T is mean air temperature in °C, and U_2 is wind speed in m s^{-1} at 2 m above the ground [and RH or dew point and air temperature are assumed to be measured at 2 m above the ground, also] provide procedures for estimating all the parameters consistent with for a grass reference crop with the defined hypothetical characteristics. The data required are the daily solar irradiance, daily maximum and minimum air temperature, mean daily dew point temperature (or daily maximum and minimum RH), mean daily wind speed at 2-m elevation and the site elevation, latitude, and longitude. Eqn. 13 can be applied using hourly data if the constant value "900" is divided by 24 for the hours in a day and the R_n and G terms are expressed as MJ m^{-2} hr^{-1}. Allen used eqn. 13 on an hourly basis in Utah with success, particularly if they corrected the aerodynamic resistance for atmospheric stability even with a constant r_s (~70 s m^{-1}) throughout the day and night.

The ASCE-EWRI Standardized Penman-Monteith Equation

In 1999, the ASCE Environmental and Water Resources Institute Evapotranspiration in Irrigation and Hydrology Committee was asked by the Irrigation Association to propose one standardized equation and set of procedures for estimating the parameters to gain consistency and wider acceptance of ET models. This committee formed a Task Committee chaired by Ivan Walter that held a series of meetings and vigorous debates on means to standardize the reference ET computations using the Penman-Monteith equation. The Task Committee built on the FAO-56 frame to develop reference ET computations that could be based on the latest engineering and scientific principles and that could be defended and that would be accurate and applicable across diverse climates. The committee had diverse geographic representation and diverse disciplines represented. The principle outcome was that TWO equations (one for a short crop named ET_{os} and another for a taller crop named ET_{rs}) were developed for daily (24 hr) and hourly (or even shorter) time periods. Allen outlined the purpose and needs for a standardized reference ET methodology.

The ASCE-EWRI standardized reference ET equation based on the FAO-56 Penman Monteith equation for a hypothetical crop with typical characteristics given in Table below is given as

$$ET_{sz} = \frac{0.408\Delta(R_n - G) + \gamma \dfrac{C_n}{T+273} U_2(e_s{}^\circ - e_a)}{\Delta + \gamma\left(1 + C_d U_2\right)}$$

where ET_{sz} is the standardized reference crop evapotranspiration for a short reference crop (ET_{os}) or a tall reference crop (ET_{rs}) in units based on the time step of mm d^{-1} for a 24-hr day or mm hr^{-1} for an hourly time step [time units on R_n and G match those for the evapotranspiration rates], C_n is the numerator constant for the reference crop type and time step, and C_d is the denominator constant for the reference crop type and time step. The ASCE-EWRI Standardized reference ET manual provides a more thorough derivation of procedures for estimating R_n and G for both reference crops on hourly time steps beyond the FAO-56 book. The ASCE-EWRI manual also addresses important issues on table below. Reference crop characteristics and Penman-Monteith equation constants for the standardized ASCE-EWRI equation.

Term	ET_{os} (short reference crop)	ET_{rs} (tall reference crop)
Vegetation height, h_c	0.12 m	0.5 m
Height of wind speed measurement, z_w	2 m	2 m
Height of air temperature and RH measurements, z_r	1.5 – 2.5 m	1.5 – 2.5 m
Zero-plane displacement height, d	0.08 m	0.08 m§
Latent heat of vaporization, λ	2.45 MJ kg^{-1}	2.45 MJ kg^{-1}
Surface resistance, r_s, daily	70 s m^{-1}	45 s m^{-1}
Surface resistance, r_s, daytime	50 s m^{-1}	30 s m^{-1}
Surface resistance, r_s, nighttime	200 s m^{-1}	200 s m^{-1}
R_n cutoff for daytime	> 0 MJ m^{-2} hr^{-1}	> 0 MJ m^{-2} hr^{-1}
R_n cutoff for nighttime	≤ 0 MJ m^{-2} hr^{-1}	≤ 0 MJ m^{-2} hr^{-1}

§ The zero-plane displacement height for ETrs assumes U2 is measured over clipped grass.

Table: Values of C_n and C_d for eqn. ($ET_{sz} = \dfrac{0.408\Delta(R_n - G) + \gamma \dfrac{C_n}{T+273} U_2(e_s{}^\circ - e_a)}{\Delta + \gamma\left(1 + C_d U_2\right)}$).

Calculation Time step	Short Reference Crop, ET_{os}		Tall Reference Crop, ET_{rs}		Units for ET_{os}, ET_{rs}	Units for R_n and G
	C_n	C_d	C_n	Cd		
Daily	900	0.34	1600	0.38	mm d^{-1}	MJ m^{-2} d^{-1}
Hourly, daytime	37	0.24	66	0.25	mm hr^{-1}	MJ m^{-2} hr^{-1}
Hourly, nighttime	37	0.96	66	1.7	mm hr^{-1}	MJ m^{-2} hr^{-1}

weather data quality assurance and estimating missing climatic data needed in the Penman-Monteith equation.

References

- Allen, R.G.; Pereira, L.S.; Raes, D.; Smith, M. (1998). Crop Evapotranspiration: Guidelines for Computing Crop Water Requirements. FAO Irrigation and drainage paper 56. Rome, Italy: Food and Agriculture Organization of the United Nations. ISBN 92-5-104219-5.

- Evaporation, encyclopedia: nationalgeographic.org, Retrieved 19 July 2018

- What-is-evaporation-and-why-is-it-important: worldatlas.com, Retrieved 10 May 2018

- Nouri, Hamideh; Beecham, Simon; Kazemi, Fatemeh; Hassanli, Ali Morad (2013). "A review of ET measurement techniques for estimating the water requirements of urban landscape vegetation". Urban Water J. 10 (4): 247–259. doi:10.1080/1573062X.2012.726360.

- Evaporation-and-factors-affecting-it, matter-in-our-surroundings, chemistry: toppr.com, Retrieved 19 March 2018

- Jasechko, Scott; Sharp, Zachary D.; Gibson, John J.; Birks, S. Jean; Yi, Yi; Fawcett, Peter J. (3 April 2013). "Terrestrial water fluxes dominated by transpiration". Nature. 496 (7445): 347–50. doi:10.1038/nature11983. PMID 23552893. Retrieved 4 April 2013.

- What-is-pan-evaporation: worldatlas.com, Retrieved 26 June 2018

- Evapotranspiration-1580: maximumyield.com, Retrieved 14 May 2018

Understanding Surface Runoff

The flow of excess meltwater and stormwater over the surface of the Earth is known as surface runoff. It is a significant component of the hydrological cycle. This chapter discusses diverse aspects of surface runoff and the various processes of snowmelt, rainfall runoff, stormwater runoff, urban runoff, etc.

If the amount of water falling on the ground is greater than the infiltration rate of the surface, runoff or overland flow will occur. Runoff specifically refers to the water leaving an area of drainage and flowing across the land surface to points of lower elevation. It is not the water flowing beneath the surface of the ground. This type of water flow is called through flow. Runoff involves the following events:

- Rainfall intensity exceeds the soil's infiltration rate.

- A thin water layer forms that begins to move because of the influence of slope and gravity.

- Flowing water accumulates in depressions.

- Depressions overflow and form small rills.

- Rills merge to form larger streams and rivers.

- Streams and rivers then flow into lakes or oceans.

On a global scale, runoff occurs because of the imbalance between evaporation and precipitation over the Earth›s land and ocean surfaces.. In fact, 86% of the Earth's evaporation occurs over the oceans, while only 14% occurs over land. Of the total amount of water evaporated into the atmosphere, precipitation returns only 79% to the oceans, and 21% to the land. Surface runoff sends 7% of the land based precipitation back to the ocean to balance the processes of evaporation and precipitation.

The distribution of runoff per continent shows some interesting patterns. Areas having the most runoff are those with high rates of precipitation and low rates of evaporation.

Table: Continental runoff

Continent	Runoff Per Unit Area (mm per yr.)
Europe	300
Asia	286

Africa	139
North and Central America	265
South America	445
Australia, New Zealand and New Guinea	218
Antarctica and Greenland	164

Streamflow and Stream Discharge

The term streamflow describes the process of water flowing in the organized channels of a stream or river. Stream discharge represents the volume of water passing through a river channel during a certain period of time. Stream discharge can be expressed mathematically with the following equation:

$$Q = W \times D \times V$$

where,

Q equals stream discharge usually measured in cubic meters per second, W equals channel width, D equals channel depth, and V equals velocity of flowing water.

Because of streamflow's potential hazard to humans many streams are gauged by mechanical recorders. These instruments record the stream's discharge on a hydrograph. The graph below illustrates a typical hydrograph and its measurement of discharge over time.

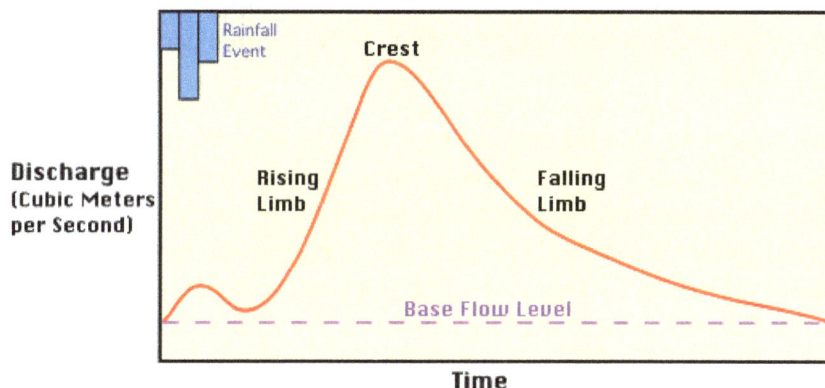

Figure: Stream hydrograph.

From this graph we can observe the following things:

- A small blip caused by rain falling directly into the channel is the first evidence that stream discharge is changing because of the rainfall.

- A significant time interval occurs between the start of rain and the beginning of the main rise in discharge on the hydrograph. This lag occurs because of the time required for the precipitation that falls in the stream's basin to eventually reach the recording station. Usually, the larger the basin the greater the the time lag.

- The rapid movement of surface runoff into the stream›s channels and subsequent flow causes the discharge to rise quickly.

- The falling limb of the hydrograph tends to be less steep that the rise. This flow represents the water added from distant tributaries and from through flow that occurs in surface soils and sediments.

- After some time the hydrograph settles at a constant level known as base flow stage. Most of the base flow comes from groundwater flow which moves water into the stream channel very slowly.

First Flush

The "first flush" phenomenon is generally assumed for rainfall events, and can be described as a concentration first flush or a mass first flush. A concentration first flush occurs when the first run-off has high concentration relative to runoff later in the storm event. A mass first (concentration times flow rate) is flow dependent and it will occurs when both concentration and the initial runoff is high relative to mass emission rate in the later runoff. Concentration first flushes have been frequently reported, but mass first flushes have rarely been quantified. For example, most of the parameters monitored for all the events in this study had higher concentrations at the beginning of the runoff than later in the runoff. Mass first flushes were less frequently observed and with lower magnitudes. This is due to the nature of the runoff, which generally has lower flow rate at the beginning of the storm than in the middle of the storm. Therefore, the mass emission rate in the middle of the storm event may be greater than at the beginning of the storm event, in spite of lower concentrations in the middle of the storm. The concept can be applied to any particular constituent or water quality parameter. Therefore, a first flush in total organic carbon (TOC), for example, can be called a TOC first flush.

The concept of first flush can also be applied to a rainfall season. In California and many other areas of the world, rainfall occurs over distinct periods. For example, the bulk of the rainfall in Los Angeles occurs from approximately November to March, with the months of January and February usually having the greatest rainfall. The long dry period from April or May to October allows contaminants to build up. The first large rainfall of the season, occurring any time from October to January, generally mobilizes the built-up contaminants, creating a larger discharge. This phenomenon is called a "seasonal first flush."

First Flush Effect

The term "first flush effect" refers to rapid changes in water quality (pollutant concentration or load) that occur after early season rains. Soil and vegetation particles wash into streams; sediments and other accumulated organic particles on the river bed are re-suspended, and dissolved substances from soil and shallow groundwater can be flushed into streams. Recent research has shown that this effect has not been observed in relatively pervious areas.

The term is often also used to address the first flood after a dry period, which is supposed to contain higher concentrations than a subsequent one. This is referred to as "first flush flood." There are various definitions of the first flush phenomenon.

First Foul Flush

Storm water runoff in a combined sewer produces a first foul flush with a suspension of accumulated sanitary solids from the sewer in addition to pollutants from surface runoff. Inflow may produce a foul flush effect in sanitary sewers if flows peak during wet weather. As flow rates increase above average, a relatively small percentage of the total flow contains a disproportionately large percentage of the total pollutant mass associated with overall flow volume through the peak flow event. Sewer solids deposition during low flow periods and subsequent resuspension during peak flow events is the major pollutant source for the first-flush combined-sewer overflow (CSO) phenomenon.

Sanitary sewage solids can either go through the system or settle out in laminar flow portions of the sewer to be available for washout during peak flows. The wetted perimeter of sewers may also be colonized by biofilm nourished by soluble sanitary wastes. Hydraulic design is the underlying reason for solids deposition in sewers. Combined sewers sized for peak runoff events expected once a decade can carry up to 1,000 times the average sanitary flow. Less dramatically oversized sewers are common in new developments and near the upstream end of collection systems. Suspended solids may accumulate when low-flow fluid velocities generate insufficient turbulence. Solids deposition is greatest where velocities are low during dry weather. In large combined sewers it may be impossible to attain sanitary sewage velocities generating sufficient turbulence to keep solids suspended during dry weather.

Biofilm and previously deposited solids may be scoured and re-entrained during peak flow turbulence.[1] The high pollution load in wastewater at the beginning of a runoff event occurs when increased flow rate erodes accumulated sewer sediment. Erosion of sediments in sewers can release pollutants in concentrations exceeding levels found in contributing sources. The initial highly polluting foul flush is released at the start of wet weather flow during speedy erosion of a weak layer of highly concentrated surficial sediment bed-load. When conditions favor dry-weather solids deposition, the first foul flush may contain as much as 30 percent of the annual total suspended solids discharged to a combined sewer system. Combined sewer suspended solids concentrations of several thousand milligrams per liter (mg/L) may be observed during the first foul flush.

Pollutant concentration levels are influenced by the age and condition of the collection system and the amount of infiltration/inflow in comparison to the sanitary flow. Pollutant concentration peaks depend on size and slope of the piping system, time interval between storms, and solids accumulation in the collection system. Steeper sewer gradients and pipe bottom shapes that maintain high velocity flow during low-flow conditions will reduce sediment accumulation in sewers; and periodic sewer flushing of individual lines during dry weather may move accumulated solids to the wastewater treatment plant before stormwater runoff causes simultaneous peak flow in the entire collection system.

Meltwater Runoff

As air temperatures rise in the spring, snowpack begins to melt. The water that becomes runoff into streams or recharge into soil is called snowmelt. Most of the snowmelt is a source of water for groundwater recharge. Snowmelt contributes substantial water to soil moisture. Part of the snowmelt that infiltrates into the soil contributes to shallow and deep groundwater. In the late spring the water levels in the river are high because of heavy rainfall experienced at this time. Once rainfall becomes less frequent in the fall, the snowmelt that was stored in the ground will return to the surface and make up the baseflow of the river. In mountains, snowmelt primarily occurs from May to July. Snowmelt occurs earlier in the prairie regions.

Snowmelt

As air temperatures rise in the spring, snowpack begins to melt. The water that becomes runoff into streams or recharge into soil is called snowmelt. Most of the snowmelt is a source of water for groundwater recharge. Snowmelt contributes substantial water to soil moisture. Part of the snowmelt that infiltrates into the soil contributes to shallow and deep groundwater. In the late spring the water levels in the river are high because of heavy rainfall experienced at this time. Once rainfall becomes less frequent in the fall, the snowmelt that was stored in the ground will return to the surface and make up the baseflow of the river. In mountains, snowmelt primarily occurs from May to July. Snowmelt occurs earlier in the prairie regions.

Figure: Snowmelt. When snowpack melts in the spring, some of the snowmelt enters streams, increasing the amount of water in the stream

Snowmelt and Flooding

The effect of snowmelt on potential flooding, mainly during the spring, is something that causes concern for many people around the world. Besides flooding, rapid snowmelt can trigger landslides and debris flows. In alpine regions like Switzerland, snowmelt is a major component of runoff. In combination with specific weather conditions, such as excessive rainfall on melting snow for example, it may even be a major cause of floods. In Switzerland, snowmelt forecasting is being used as a flood-warning tool to predict snowmelt runoff and potential flooding.

In some parts of the world, such as in the Pacific Northwest of the United States, annual springtime flood events occur when rain falls on existing snowpacks, known as a "rain-on-snow event." Runoff during rain-on-snow events has been associated with mass-wasting of hill slopes, damage to riparian (areas alongside streams) zones, downstream flooding and associated damage, and loss of life. Some studies suggest that the amount of forest cover can have an influence on the magnitude of rain-on-snow events.

Hetch-Hetchy basin near Yosemite, California. Photo by David Gay

Figure: Snowmelt in the Hetch-Hetchy basin near Yosemite, California

Conodoguinet Creek, Pennsylvania. NOAA, Jason Nolan

Figure: Snowmelt in the Conodoguinet creek, Pennsylvania

Rainfall Runoff

Runoff is generated by rainstorms and its occurrence and quantity are dependent on the characteristics of the rainfall event, i.e. intensity, duration and distribution. There are, in addition, other important factors which influence the runoff generating process.

Rainfall Characteristics

Precipitation in arid and semi-arid zones results largely from convective cloud mechanisms producing storms typically of short duration, relatively high intensity and limited areal extent. However, low intensity frontal-type rains are also experienced, usually in the winter season. When most precipitation occurs during winter, as in Jordan and in the Negev, relatively low-intensity rainfall may represent the greater part of annual rainfall.

Rainfall intensity is defined as the ratio of the total amount of rain (rainfall depth) falling during a given period to the duration of the period It is expressed in depth units per unit time, usually as mm per hour (mm/h).

The statistical characteristics of high-intensity, short-duration, convective rainfall are essentially independent of locations within a region and are similar in many parts of the world. Analysis of short-term rainfall data suggests that there is a reasonably stable relationship governing the intensity characteristics of this type of rainfall. Studies carried out in Saudi Arabia (Raikes and Partners 1971) suggest that, on average, around 50 percent of all rain occurs at intensities in excess of 20 mm/hour and 20-30 percent occurs at intensities in excess of 40 mm/hour. This relationship appears to be independent of the long-term average rainfall at a particular location.

Variability of Annual Rainfall

Water harvesting planning and management in arid and semi-arid zones present difficulties which are due less to the limited amount of rainfall than to the inherent degree of variability associated with it.

In temperate climates, the standard deviation of annual rainfall is about 10-20 percent and in 13 years out of 20, annual amounts are between 75 and 125 percent of the mean. In arid and semi-arid climates the ratio of maximum to minimum annual amounts is much greater and the annual rainfall distribution becomes increasingly skewed with increasing aridity. With mean annual rainfalls of 200-300 mm the rainfall in 19 years out of 20 typically ranges from 40 to 200 percent of the mean and for 100 mm/year, 30 to 350 percent of the mean. At more arid locations it is not uncommon to experience several consecutive years with no rainfall.

For a water harvesting planner, the most difficult task is therefore to select the appropriate "design" rainfall according to which the ratio of catchment to cultivated area will be determined.

Design rainfall is defined as the total amount of rain during the cropping season at which or above which the catchment area will provide sufficient runoff to satisfy the crop water requirements. If the actual rainfall in the cropping season is below the design rainfall, there will be moisture stress in the plants; if the actual rainfall exceeds the design rainfall, there will be surplus runoff which may result in a damage to the structures.

The design rainfall is usually assigned to a certain probability of occurrence or exceedance. If, for example, the design rainfall with a 67 percent probability of exceedance is selected, this means that on average this value will be reached or exceeded in two years out of three and therefore the crop water requirements would also be met in two years out of three.

The design rainfall is determined by means of a statistical probability analysis.

Probability Analysis

The first step is to obtain annual rainfall totals for the cropping season from the area of concern. In locations where rainfall records do not exist, figures from stations nearby may be used with caution. It is important to obtain long-term records the variability of rainfall in arid and semi-arid areas is considerable. An analysis of only 5 or 6 years of observations is inadequate as these 5 or 6 values may belong to a particularly dry or wet period and hence may not be representative for the long term rainfall pattern.

In the following example, 32 annual rainfall totals from Mogadishu (Somalia) were used for an analysis.

Table: annual rainfall, mogadishu (somalia)

Year	R mm	Year	R mm	Year	R mm	Year	R mm	Year	R mm
1957	484	1964	489	1971	271	1977	660	1983	273
1958	529	1965	498	1972	655	1978	216	1984	270
1959	302	1966	395	1973	371	1979	594	1985	423
1960	403	1967	890	1974	255	1980	544	1986	251
1961	960	1968	680	1975	411	1981	563	1987	533
1962	453	1969	317	1976	339	1982	526	1988	531
1963	633	1970	300						

The next step is to rank the annual totals from table above with m == 1 for the largest and m = 32 for the lowest value and to rearrange the data accordingly.

The probability of occurrence P (%) for each of the ranked observations can be calculated from the equation

$$P(\%) = \frac{m - 0.375}{N + 0.25} \times 100$$

where:

P = probability in % of the observation of the rank m
m = the rank of the observation
N = total number of observations used

Table: ranked annual rainfall data, mogadishu (somalia)

Year	R	m	P	Year	R	m	P	Year	R	m	P	Year	R	m	P
	mm		%		mm		%		mm		%		mm		%
1961	960	1	1.9	1988	531	11	32.9	1966	395	21	64.0	1986	251	31	95.0
1967	890	2	5.0	1958	529	12	36.0	1973	371	22	67.1	1978	216	32	98.1
1968	680	3	8.1	1982	526	13	39.1	1976	339	23	70.2				
1977	660	4	11.2	1965	498	14	42.2	1969	317	24	73.3				
1972	655	5	14.3	1964	489	15	45.3	1959	302	25	76.4				
1963	633	6	17.4	1957	484	16	48.4	1970	300	26	79.5				
1979	594	7	20.5	1962	453	17	51.6	1983	273	27	82.6				
1981	563	8	23.6	1985	423	18	54.7	1971	271	28	85.7				
1980	544	9	26.7	1975	411	19	57.8	1984	270	29	88.8				
1987	533	10	29.8	1960	403	20	60.9	1974	255	30	91.1				

The above equation is recommended for N = 10 to 100 There are several other, but similar, equations known to compute experimental probabilities.

The next step is to plot the ranked observations (columns 2, 6,10, 14, Table above) against the corresponding probabilities (columns 4, 8,12,16, Table above). For this purpose normal probability paper must be used.

Finally a curve is fitted to the plotted observations in such a way that the distance of observations above or below the curve should be as close as possible to the curve. The curve may be a straight line.

From this curve it is now possible to obtain the probability of occurrence or exceedance of a rainfall value of a specific magnitude. Inversely, it is also possible to obtain the magnitude of the rain corresponding to a given probability.

In the above example, the annual rainfall with a probability level of 67 percent of exceedance is 371 mm, i.e. on average in 67 percent of time (2 years out of 3) annual rain of 371 mm would be equalled or exceeded.

For a probability of exceedance of 33 percent, the corresponding value of the yearly rainfall is 531 mm.

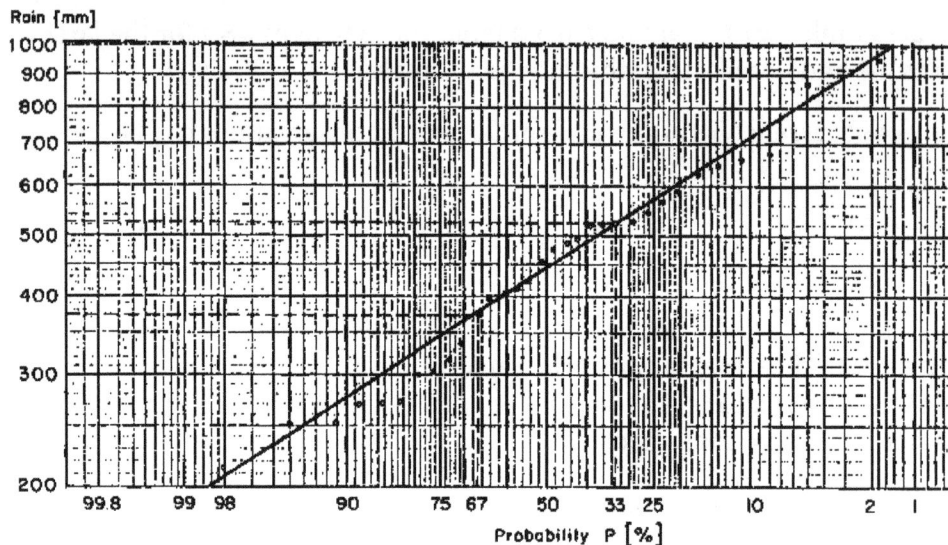

Figure: Probability diagram with regression line for an observed series of annual rainfall totals -Mogadishu, Somalia

The return period T (in years) can easily be derived once the exceedance probability P (%) is known from the equations.

$$T = \frac{100}{P}(Years)$$

From the above examples the return period for the 67 percent and the 33 percent exceedance probability events would thus be:

$$T_{67\%} = \frac{100}{67} = 1.5(Years)$$

i.e. on average an annual rainfall of 371 mm or higher can be expected in 2 years out of 3;

$$T_{33\%} = \frac{100}{33} = 3(Years)$$

respectively i.e. on average an annual rainfall of 531 mm or more can only be expected in 1 year out of 3.

Rainfall-runoff Relationship

The Surface Runoff Process

When rain falls, the first drops of water are intercepted by the leaves and stems of the vegetation. This is usually referred to as interception storage

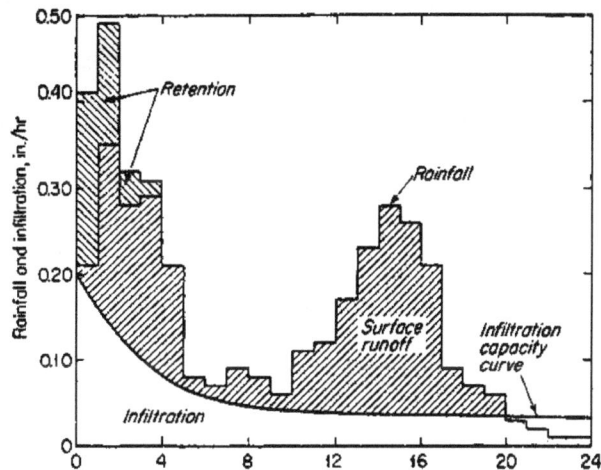

Figure: Schematic diagram illustrating relationship between rainfall, infiltration and runoff

As the rain continues, water reaching the ground surface infiltrates into the soil until it reaches a stage where the rate of rainfall (intensity) exceeds the infiltration capacity of the soil. Thereafter, surface puddles, ditches, and other depressions are filled (depression storage), after which runoff is generated.

The infiltration capacity of the soil depends on its texture and structure, as well as on the antecedent soil moisture content (previous rainfall or dry season). The initial capacity (of a dry soil) is high but, as the storm continues, it decreases until it reaches a steady value termed as final infiltration rate.

The process of runoff generation continues as long as the rainfall intensity exceeds the actual infiltration capacity of the soil but it stops as soon as the rate of rainfall drops below the actual rate of infiltration.

The rainfall runoff process is well described in the literature. Numerous papers on the subject have been published and many computer simulation models have been developed. All these models, however, require detailed knowledge of a number of factors and initial boundary conditions in a catchment area which in most cases are not readily available.

For a better understanding of the difficulties of accurately predicting the amount of runoff resulting from a rainfall event, the major factors which influence the rainfall-runoff process are described below.

Factors Affecting Runoff

Apart from rainfall characteristics such as intensity, duration and distribution, there are a number of site (or catchment) specific factors which have a direct bearing on the occurrence and volume of runoff.

Soil Type

The infiltration capacity is among others dependent on the porosity of a soil which determines the water storage capacity and affects the resistance of water to flow into deeper layers.

Porosity differs from one soil type to the other. The highest infiltration capacities are observed in loose, sandy soils while heavy clay or loamy soils have considerable smaller infiltration capacities.

Figure below illustrates the difference in infiltration capacities measured in different soil types.

The infiltration capacity depends furthermore on the moisture content prevailing in a soil at the onset of a rainstorm.

The initial high capacity decreases with time (provided the rain does not stop) until it reaches a constant value as the soil profile becomes saturated.

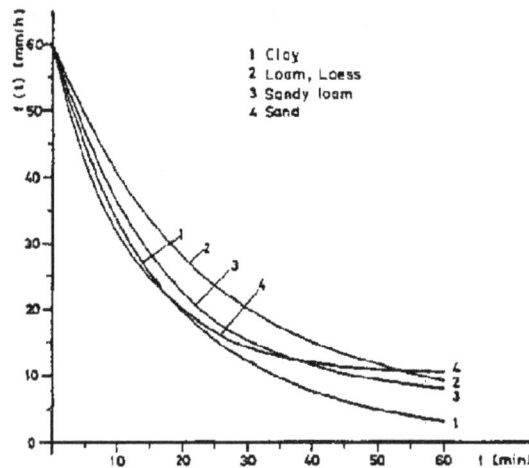

Figure: Infiltration capacity curves for different soil types

This however, is only valid when the soil surface remains undisturbed.

It is well known that the average size of raindrops increases with the intensity of a rainstorm. In a high intensity storm the kinetic energy of raindrops is considerable when hitting the soil surface. This causes a breakdown of the soil aggregate as well as soil dispersion with the consequence of driving fine soil particles into the upper soil pores. This results in clogging of the pores, formation of a thin but dense and compacted layer at the surface which highly reduces the infiltration capacity.

This effect, often referred to as capping, crusting or sealing, explains why in arid and semi-arid areas where rainstorms with high intensities are frequent, considerable quantities of surface runoff are observed even when the rainfall duration is short and the rainfall depth is comparatively small.

Soils with a high clay or loam content (e.g. Loess soils with about 20% clay) are the most sensitive for forming a cap with subsequently lower infiltration capacities. On coarse, sandy soils the capping effect is comparatively small.

Vegetation

The amount of rain lost to interception storage on the foliage depends on the kind of vegetation and its growth stage. Values of interception are between 1 and 4 mm. A cereal crop, for example, has a smaller storage capacity than a dense grass cover.

The amount of rain lost to interception storage on the foliage depends on the kind of vegetation and its growth stage. Values of interception are between 1 and 4 mm. A cereal crop, for example, has a smaller storage capacity than a dense grass cover.

More significant is the effect the vegetation has on the infiltration capacity of the soil. A dense vegetation cover shields the soil from the raindrop impact and reduces the crusting effect as described earlier.

In addition, the root system as well as organic matter in the soil increase the soil porosity thus allowing more water to infiltrate. Vegetation also retards the surface flow particularly on gentle slopes, giving the water more time to infiltrate and to evaporate.

In conclusion, an area densely covered with vegetation, yields less runoff than bare ground.

Slope and Catchment Size

Investigations on experimental runoff plots have shown that steep slope plots yield more runoff than those with gentle slopes.

In addition, it was observed that the quantity of runoff decreased with increasing slope length.

This is mainly due to lower flow velocities and subsequently a longer time of concentration (defined as the time needed for a drop of water to reach the outlet of a catchment from the most remote location in the catchment). This means that the water is exposed for a longer duration to infiltration and evaporation before it reaches the measuring point. The same applies when catchment areas of different sizes are compared.

The runoff efficiency (volume of runoff per unit of area) increases with the decreasing size of the catchment i.e. the larger the size of the catchment the larger the time of concentration and the smaller the runoff efficiency.

Figure below clearly illustrates this relationship.

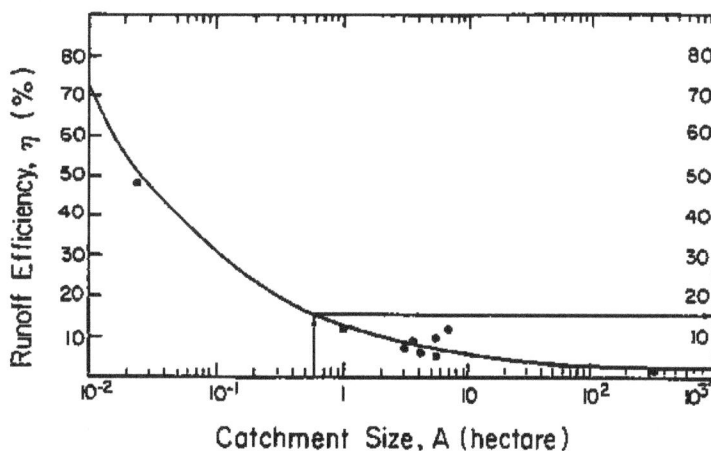

Figure: Runoff efficiency as a function of catchment size (Ben Asher 1988)

It should however be noted that the diagram has been derived from investigations in the Negev desert and not be considered as generally applicable to others regions. The purpose of this diagram is to demonstrate the general trend between runoff and catchment size.

Runoff Coefficients

Apart from the above-mentioned site-specific factors which strongly influence the rainfall-runoff process, it should also be considered that the physical conditions of a catchment area are not homogenous. Even at the micro level there are a variety of different slopes, soil types, vegetation covers etc. Each catchment has therefore its own runoff response and will respond differently to different rainstorm events.

The design of water harvesting schemes requires the knowledge of the quantity of runoff to be produced by rainstorms in a given catchment area. It is commonly assumed that the quantity (volume) of runoff is a proportion (percentage) of the rainfall depth.

Runoff [mm] = K x Rainfall depth [mm]

In rural catchments where no or only small parts of the area are impervious, the coefficient K, which describes the percentage of runoff resulting from a rainstorm, is however not a constant factor. Instead its value is highly variable and depends on the above described catchment-specific factors and on the rainstorm characteristics.

For example, in a specific catchment area with the same initial boundary condition (e.g. antecedent soil moisture), a rainstorm of 40 minutes duration with an average intensity of 30 mm/h would produce a smaller percentage of runoff than a rainstorm of only 20 minutes duration but with an average intensity of 60 mm/h although the total rainfall depth of both events were equal.

Determination of Runoff Coefficients

For reasons explained before, the use of runoff coefficients which have been derived for watersheds in other geographical locations should be avoided for the design of a water harvesting scheme. Also runoff coefficients for large watersheds should not be applied to small catchment areas.

An analysis of the rainfall-runoff relationship and subsequently an assessment of relevant runoff coefficients should best be based on actual, simultaneous measurements of both rainfall and runoff in the project area.

As explained above, the runoff coefficient from an individual rainstorm is defined as runoff divided by the corresponding rainfall both expressed as depth over catchment area (mm):

$$K = \frac{Runoff\,[mm]}{RainFall\,[mm]}$$

Actual measurements should be carried out until a representative range is obtained. Shanan and Tadmor recommend that at least 2 years should be spent to measure rainfall and runoff data before any larger construction programme starts. Such a time span would in any case be justified bearing in mind the negative demonstration effect a water harvesting project would have if the structures were seriously damaged or destroyed already during the first rainstorm because the design was based on erroneous runoff coefficients.

When plotting the runoff coefficients against the relevant rainfall depths a satisfactory correlation is usually observed.

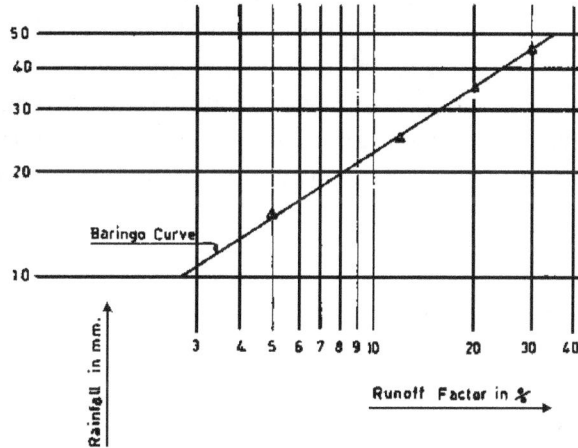

Figure: Rainfall-runoff relationships, Baringo, Kenya

A much better relationship would be obtained if in addition to rainfall depth the corresponding rainstorm intensity, the rainstorm duration and the antecedent soil moisture were also measured. This would allow rainstorm events to be grouped according to their average intensity and their antecedent soil moisture and to plot the runoff coefficients against the relevant rainfall durations separately for different intensities.

Rainfall intensities can be accurately measured by means of a continuously recording autographic rain gauge.

It is also possible to time the length of individual rainstorms and to calculate the average intensities by dividing the measured rainfall depths by the corresponding duration of the storms.

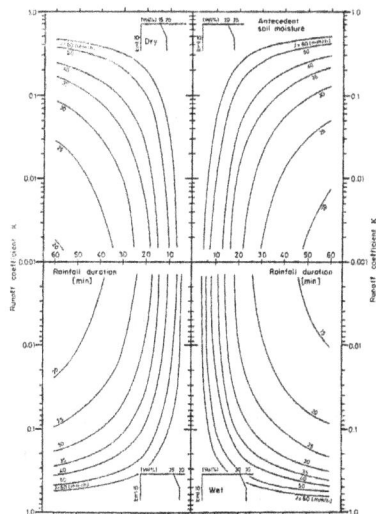

Figure: Runoff coefficients in relation to rainfall intensity, rainfall duration and antecedent soil moisture. Measured on loess soil with sparse vegetation. Ground slope 1.5%.

When analysing the measured data it will be noted that a certain amount of rainfall is always required before any runoff occurs. This amount, usually referred to as threshold rainfall, represents the initial losses due to interception and depression storage as well as to meet the initially high infiltration losses.

The threshold rainfall depends on the physical characteristics of the area and varies from catchment to catchment. In areas with only sparse vegetation and where the land is very regularly shaped, the threshold rainfall may be only in the range of 3 mm while in other catchments this value can easily exceed 12 mm, particularly where the prevailing soils have a high infiltration capacity. The fact that the threshold rainfall has first to be surpassed explains why not every rainstorm produces runoff. This is important to know when assessing the annual runoff-coefficient of a catchment area.

Assessment of Annual or Seasonal Runoff

The knowledge of runoff from individual storms as described before is essential to assess the runoff behaviour of a catchment area and to obtain an indication both of runoff-peaks which the structure of a water harvesting scheme must withstand and of the needed capacity for temporary surface storage of runoff, for example the size of an infiltration pit in a micro catchment system.

However, to determine the ratio of catchment to cultivated area, it is necessary to assess either the annual (for perennial crops) or the seasonal runoff coefficient. This is defined as the total runoff observed in a year (or season) divided by the total rainfall in the same year (or season).

$$k = \frac{Yearly(seasonal)TotalRunoff\,[mm]}{Yearly(seasonal)TotalRainFall\,[mm]}$$

The annual (seasonal) runoff coefficient differs from the runoff coefficients derived from individual storms as it takes into account also those rainfall events which did not produce any runoff. The annual (seasonal) runoff-coefficient is therefore always smaller than the arithmetic mean of runoff coefficients derived from individual runoff-producing storms.

Runoff Plots

Runoff plots are used to measure surface runoff under controlled conditions. The plots should be established directly in the project area. Their physical characteristics, such as soil type, slope and vegetation must be representative of the sites where water harvesting schemes are planned.

The size of a plot should ideally be as large as the estimated size of the catchment planned for the water harvesting project. This is not always possible mainly due to the problem of storing the accumulated runoff. A minimum size of 3-4 m in width and 10-12 m in length is recommended. Smaller dimensions should be avoided, since the results obtained from very small plots are rather misleading.

Care must be taken to avoid sites with special problems such as rills, cracks or gullies crossing the plot. These would drastically affect the results which would not be representative for the whole area. The gradient along the plot should be regular and free of local depressions. During construction of the plot, care must be taken not to disturb or change the natural conditions of the plot such as destroying the vegetation or compacting the soil. It is advisable to construct several plots in series in the project area which would permit comparison of the measured runoff volumes and to judge on the representative character of the selected plot sites.

Around the plots metal sheets or wooden planks must be driven into the soil with at least 15 cm of height above ground to stop water flowing from outside into the plot and vice versa. A rain gauge must be installed near to the plot. At the lower end of the plot a gutter is required to collect the runoff. The gutter should have a gradient of 1% towards the collection tank. The soil around the gutter should be backfilled and compacted. The joint between the gutter and the lower side of the plot may be cemented to form an apron in order to allow a smooth flow of water from the plot into the gutter. The collection tank may be constructed from stone masonry, brick or concrete blocks, but a buried barrel will also meet the requirements. The tank should be covered and thus be protected against evaporation and rainfall. The storage capacity of the tank depends on the size of the plot but should be large enough to collect water also from extreme rain storms. Following every storm (or every day at a specific time), the volume of water collected in the rain gauge and in the runoff tank must be measured. Thereafter the gauge and tank must be completely emptied. Any silt which may have deposited in the tank and in the gutter must be cleared.

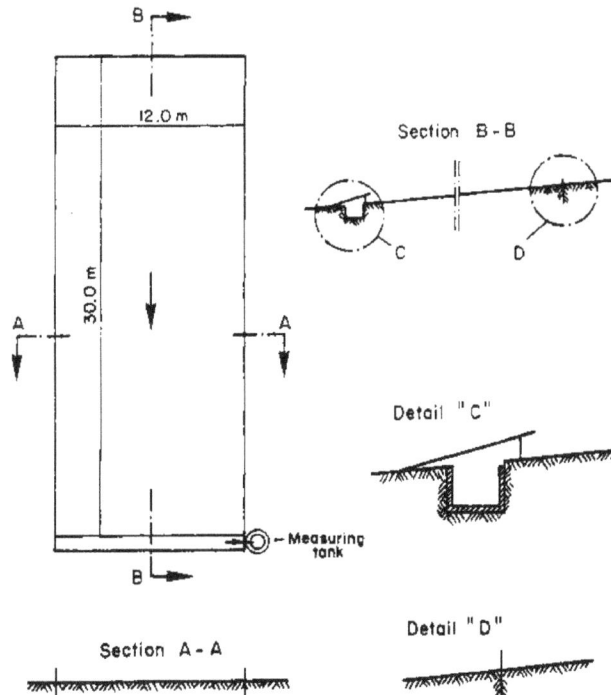

Figure: Standard layout of a runoff plot

Rainfall-Runoff Modeling

Rainfall-Runoff modeling is one of the most classical applications of hydrology. It has the purpose of simulating the peak river flow or the hydrograph induced by an observed or a hypothetical rainfall forcing. Rainfall-runoff models may include other input variables, like temperature, information on the catchment or others. Within the context of this subject, we are studying rainfall-runoff models with the purpose of producing estimates of peak river flow, simulation of flood hydrographs or simulation of synthetic river flows in general, even for extended periods, for example for setting up water resources management strategies.

Therefore, rainfall-runoff modelling is a cross cutting topic over several of the major issue this subject is focusing on. In view of its central role, rainfall-runoff modelling is then treated separately in this web page.

Rainfall-runoff models describe a portion of the water cycle and therefore the movement of a fluid - water - and therefore they are explicitly or implicitly based on the laws of physics, and in particular on the principles of conservation of mass, conservation of energy and conservation of momentum. Depending on their complexity, models can also simulate the dynamics of water quality, ecosystems, and other dynamical systems related to water, therefore embedding laws of chemistry, ecology, social sciences and so forth.

Models are built by constitutive equations, namely, mathematical formulations of the above laws, whose number depends on the number of variables to be simulated. The latter are the output variables, and the state variables, which one may need to, introduce to describe the state of the system. Constitutive equations may include parameters: they are numeric factors in the model equations that can assume different values therefore making the model flexible. In order to apply the model, parameters needs to be estimated (or calibrated, or optimized, and we say that the model is calibrated, parameterized, optimized). Parameters usually assume fixed value, but in some models they may depend on time, or the state of the system.

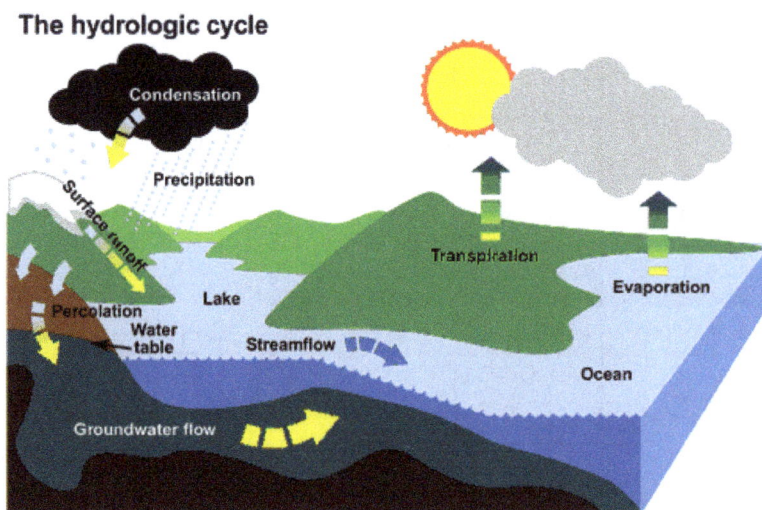

Figure: The hydrologic cycle

Rainfall-runoff models can be classified within several different categories. They can distinguished between event-based and continuous-simulation models, black-box versus conceptual versus process based (or physically based) models, lumped versus distributed models, and several others. It is important to note that the above classifications are not rigid - sometimes a model cannot be unequivocally assigned to one category. We will treat rainfall-runoff models by taking into consideration models of increasing complexity.

The Rational Formula

The Rational equation is the simplest method to determine peak discharge from drainage basin runoff. It but is the most common method used for sizing sewer systems.

Rational Equation: Q=ciA

The Rational equation requires the following units:

> Q = Peak discharge, cfs
> c = Rational method runoff coefficient
> i = Rainfall intensity, inch/hour
> A = Drainage area, acre

Note that our calculation allows you to use a variety of units.

The Rational method runoff coefficient (c) is a function of the soil type and drainage basin slope.

The Linear Reservoir

The linear reservoir is one of the most used rainfall-runoff models, together with the time area method. The linear reservoir model assimilates the catchment to a reservoir, for which the conservation of mass applies. The reservoir is fed by rainfall, and releases the river flow through a bottom discharge, for which a linear dependence applies between the river flow and the volume of water stored in the reservoir, while other losses - including evapotranspiration - are neglected. Therefore, the model is constituted by the following relationships

$$\frac{dW(t)}{dt} = p(t) - q(t)$$

and

$$W(t) = k\, q(t)$$

where W(t) is the volume of water stored in the catchment at time t, p(t) is rainfall, q(t) is the river flow at time t and k is a constant parameter with the dimension of time (if the parameter was not constant the model would not be linear).

The second equation above assumes a linear relationship between discharge and storage into the catchment. The properties of a linear function are described here. Actually, the relationship between storage in a real tank and bottom discharge is an energy conservation equation that is not linear; in fact, as it is given by the well-known Torricelli's law. Therefore the linearity assumption is an approximation, which is equivalent to assuming that the superposition principle applies to runoff generation. Actually, such assumption does not hold in practice, as the catchment response induced by two subsequent rainfall events cannot be considered equivalent to the sum of the individual catchment responses to each single event. However, linearity is a convenient assumption to make the model simpler and analytical integration possible.

A nice feature of the linear reservoir is that the above equations can be integrated analytically, under simplifying assumption. In fact, by substituting the second equation into the first one gets:

$$k\frac{dq(t)}{dt} = p(t) - q(t)$$

Then, by multiplying both sides by $e^{t/k}$ and dividing by k one gets

$$e^{t/k}\frac{dq(t)}{dt}+e^{t/k}\frac{q(t)}{k}-e^{t/k}\frac{p(t)}{k}=0$$

which can be written as:

$$\frac{d}{dt}\left[e^{t/k}q(t)\right]=e^{t/k}\frac{p(t)}{k}$$

By integrating between 0 and t one obtains:

$$\int_0^t\frac{d}{d\tau}\left[e^{\tau/k}q(\tau)\right]d\tau=\int_0^t e^{\tau/k}\frac{p(\tau)}{k}d\tau$$

$$\left[e^{\tau/k}q(\tau)\right]_0^t=\int_0^t e^{\tau/k}\frac{p(\tau)}{k}d\tau$$

$$q(t)e^{t/k}-q(0)=\int_0^t e^{\tau/k}\frac{p(\tau)}{k}d\tau$$

and, by assuming q(0)=0, one finally gets:

$$q(t)=\int_0^t\frac{1}{k}e^{\frac{-(t-\tau)}{k}}p(\tau)d\tau$$

which can be easily integrated numerically by using the Euler method.

If one assumes p(t)=constant, an explicit expression is readily obtained for the river flow:

$$q(t)=\frac{p}{k}\int_0^t e^{-\frac{t-\tau}{k}}d\tau=\frac{p}{k}e^{-\frac{t}{k}}\int_0^t e^{\frac{\tau}{k}}d\tau=\frac{p}{k}e^{-\frac{t}{k}}\left[ke^{\frac{t}{k}}\right]_0^t=\frac{p}{k}e^{-\frac{t}{k}}\left[ke^{\frac{t}{k}}-k\right]=p\left[1-e^{-\frac{t}{k}}\right]$$

The parameter k is usually calibrated by matching observed and simulated river flows. Its value significantly impacts the catchment response. A high k value implies a large storage into the catchment. Therefore, large values of k are appropriate for catchment with a significant storage capacity. Conversely, a low k value is appropriate for impervious basins. k is also related to the response time of the catchment. A low k implies a quick response, while slowly responding basins are characterized by a large k. In fact, k is related to the response time of the catchment.

The non-linear Reservoir

Several variants of the linear reservoir modeling scheme can be introduced, for instance by adopting a non linear relationship between discharge and storage. Moreover, an upper limit can be fixed for the storage in the catchment, and additional discharges can be introduced, which can be activated for different levels of storage. All of the above modifications make the model non-linear so that an analytical integration is generally not possible. One should also take into account that increasing the number of parameters implies a corresponding increase of estimation variance and therefore simulation uncertainty. Furthermore, it is often observed that introducing additional discharges or thresholds may induce discontinuities in the hydrograph shape.

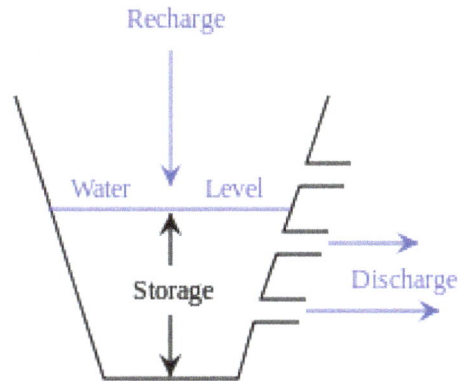

Figure: A non-linear reservoir

The Hymod Model

The Hymod model is a flexible solution that is increasingly adopted for its capability of providing a good fit in several practical applications. It was originally proposed by Boyle (2000). It is based on the assumption that each point location i in the basin is characterised by a local value of soil water storage Ci, which varies from 0 in the impervious areas up to a maximum value Cmax in the most permeable location of the catchment. Ci is assumed to be randomly varying, so that for an assigned value C* of soil water storage a probability distribution is introduced that gives the probability that a randomly selected location j is characterised by Cj less then, or equal to, C*. Such probability may be interpreted as the fraction F(C*) of the catchment area where Cj≤C*. The above probability distribution is written as

$$F(c_*) = 1 - \left(1 - \frac{c_*}{c_{\max}}\right)^{\beta k}, \quad 0 \le c_* \le c_{\max}$$

Here, βk is a parameter which quantifies the variability of the soil water storage over the catchment. One can easily verify by numerical simulation that βk=0 implies that the soil water storage is constant over the basin and equal to c_{max}; $\beta k = 1$ implies that the soil water storage is linearly varying from 0 to Cmax; βk→∞ implies that the soil water storage is tending to the null value over the whole catchment, which is therefore impervious.

Let us assume that a storm event occurs over the basin and let us define with the symbol C(t) the time varying water depth stored in the unsaturated locations of the catchment. If we ignore any water losses, like evapotranspiration, C(t) is equal to the rainfall amount from the beginning of the event. If one assumes that the shape of the above probability distribution, now expressed in terms of C(t), is the one reported in figure below, it can be easily proved that the water volume stored in the catchment at time t is given by

$$w(t) = c(t) - \left[\int_0^{c(t)} f(\chi) d\chi \right]$$

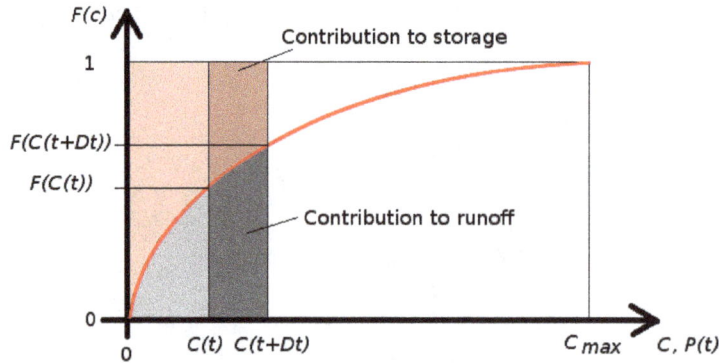

Figure: Distribution of soil water storage, surface runoff and stored volume in Hymod

In fact, the integral at the right hand side of the above equation is the area below the red line in figure above. Elementary increment of that area are given by the product of the rainfall at each time step by the fraction F(c(t)) of saturated area at the same time, namely, the product of rainfall by saturated area, which is indeed the surface runoff. Conversely, the area above the curve gives the global storage into the catchment W(t), which can be interpreted as a weighted average value of C(t). The progress of surface runoff and water storage is depicted by the animated picture in figure below. Note that after saturation, the contribution of surface runoff is given by the product of the rainfall itself by 1, which is the fraction of saturated area, taking and keeping unit value when the catchment is saturated. After saturation, the storage in the catchment of course does not increase any more.

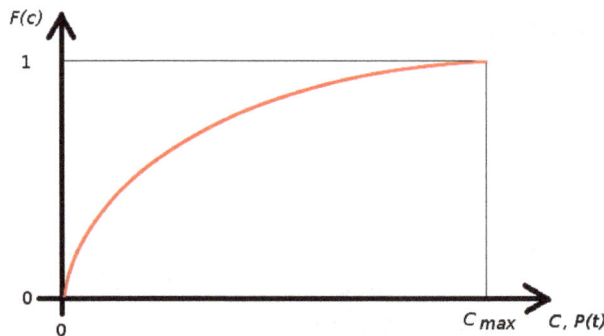

Figure: Progress of surface runoff and stored volume in Hymod

By computing the integral in the latter relationship one obtains:

$$\int_0^{C(t)} F(\chi) dx = \int_0^{C(t)} \left\{ 1 - \left[1 - \frac{\chi}{C_{max}} \right]^{\beta_k} \right\} d\chi$$

$$\int_0^{C(t)} F(\chi)dx = \left[\chi + \frac{C_{max}}{(\beta_k +1)}\left[1 - \frac{\chi}{C_{max}}\right]^{\beta_k +1}\right]_0^{C(t)}$$

$$\int_0^{C(t)} F(\chi)d\chi = C(t) + \frac{C_{max}}{(\beta_k +1)}\left[1 - \frac{C(t)}{C_{max}}\right]^{\beta_k +1} - \frac{C_{max}}{(\beta_k +1)}$$

Therefore the water volume stored in the catchment is equal to

$$W(t) = \frac{C_{max}}{(\beta_k +1)} - \frac{C_{max}}{(\beta_k +1)}\left[1 - \frac{C(t)}{C_{max}}\right]^{\beta_k +1} = \frac{C_{max}}{(\beta_k +1)}\left\{1 - \left[1 - \frac{C(t)}{C_{max}}\right]^{\beta_k +1}\right\}$$

By inverting the above equation one gets

$$C(t) = C_{max}\left[1 - \left(1 - w(t)\frac{\beta_k +1}{c_{max}}\right)^{\frac{1}{\beta_k +1}}\right]$$

Note that one obtains an estimate of the upper value of the water volume that the catchment can store by imposing C(t)=Cmax, therefore obtaining

$$W_{max} = \frac{C_{max}}{(\beta_k +1)}$$

The above equations allow an easy application of the Hymod model through a numerical simulation, that is usually carried out by adopting a time step Dt that is equal to observational time step of rainfall and river flow. At a given time step t, one knows the value of C(t) which is equal to the cumulative rainfall depth from the beginning of the event at time t. Therefore, W(t) can be easily computed as well by using the above relationships. At the time t+1, C(t+1)=C(t)+P(t), where P is rainfall, under the condition that C(t+1)=Cmax if C(t)+P(t)>Cmax. Therefore, one can compute a first contribution to surface runoff through the relationship

$$ER_1(t)=max(C(t)+P(t)-Cmax,0).$$

Finally, one can compute a second contribution to the surface runoff which is given by the water volume that cannot be absorbed by the catchment because part of the catchment area got saturated in the last time step. Such second contribution is given by

$$ER_2(t)=(C(t+1)-C(t))-(W(t+1)-W(t)).$$

By summing ER1+ER2 one obtains the total contribution to the surface runoff during the time step from t to t+1.

At this stage, one may evaluate the water losses given by evapotranspiration at the current time step, and subtract them from W(t+1). Such water losses are computed within Hymod through the relationship

$$E(t) = \left(1 - \frac{\frac{C_{max}}{\beta_k +1} - W(t)}{\frac{C_{max}}{\beta_k +1}}\right)E_p(t)$$

Here, Ep(t) is the potential evapotranspiration at time t. Then, the water storage at time t+1 is given by

$$W(t+1)=W(t)-Ep(t).$$

One should note that the evapotranspiration is subtracted from the stored water volume after ER1 and ER2 are computed.

The total contribution ER(t)=ER1(t)+ER2(t) to the surface runoff is then divided into 2 components: αER(t) which represent the fast runoff and (1-α)ER(t) which is the slow runoff. αER(t) is propagated through a series of linear reservoirs with the same bottom discharge time constant kq, while (1-α)ER(t) is instead propagated through a single linear reservoir with parameter ks.

The computation moves forward through the sequence of time steps. Hymod counts 5 parameters, namely: Cmax, β, α, kq and ks. These parameters need to be calibrated by using observed data.

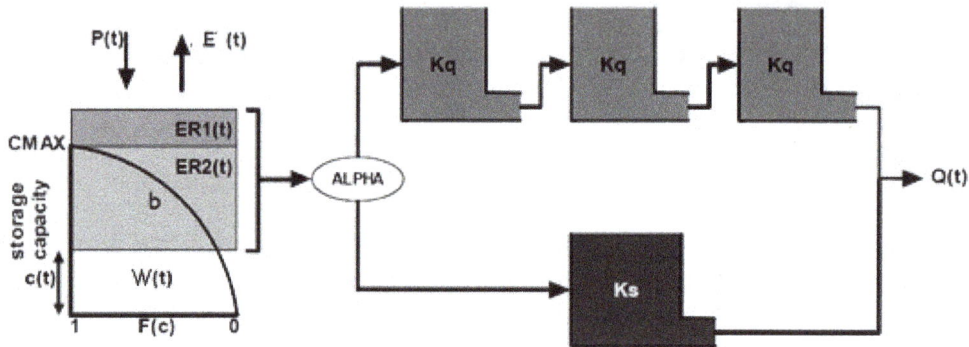

Figure: A schematic representation of the Hymod model

Stormwater Runoff

Stormwater runoff occurs after a rainfall. Storm water flows over impervious (unable to penetrate) surfaces like driveways, sidewalks, streets, parking lots and roofs and is unable to percolate (filter or seep) into the ground. This unfiltered water reaches our neighborhood streams, ponds, lakes, bays, wetlands and oceans and can eventually make its way into our ground water.

Affect of Pollution on Watershed

There are two basic types of pollution: point source and non-point source. Point source pollution is easy to understand because it can be traced directly to its source. Point source pollution was a big concern in the past, but today stricter laws and regulations have drastically decreased the problem.

Non-point source pollution is a little more difficult to understand. Stormwater runoff pollution is a type of non-point source pollution. This means that the pollution cannot be traced back to a specific source, but instead comes from many different sources throughout the environment. Non-point source pollution is the primary cause of watershed pollution today. Non-point source pollution occurs when small amount of pollution from a large variety of sources is picked up by stormwater runoff and carried into water bodies.

Land cover affects

Stormwater runoff can carry many different types of non-point source pollution. Each can affect your watershed in a different way. Sediment (dirt, soil, sand) can increase the turbidity (a measure of water cloudiness) of a water body. Turbidity can block sunlight from reaching aquatic plants, making it impossible for them to grow. Without plants, animals lose a food source and it is more difficult to filter pollutants from the water. Instead, pollutants collect in the bottom of the water body and remain there indefinitely.

Excess nutrients carried in stormwater runoff can also negatively affect our water supply. These nutrients, primarily nitrogen and phosphorus, can come from lawn fertilizers or natural sources, such as manure. Nutrients can cause algal and bacterial blooms, which proliferate (reproduce) rapidly. Algae will consume oxygen, increase turbidity in the water body and eventually die along with the fish and other aquatic life that need oxygen to live.

Debris such as plastic bags, bottles and cigarette butts can wash into a water body and interfere with aquatic life. Other hazardous wastes can be carried into a water body. These include insecticides, (chemicals used to control or kill insects) herbicides, (chemicals used to kill unwanted plants) paint, motor oil and heavy metals.

Stormwater Runoff: A Problem

Stormwater washes pollutants off roads, lawns and other surfaces and carries them into the nearest body of water. Some of the pollutants commonly carried by runoff are:

- Sediment

- Oil, grease, and toxic chemicals from motor vehicles

- Pesticides and nutrients from lawns and gardens

- Viruses, bacteria, and nutrients from pet waste and failing septic systems

- Heavy metals from roof shingles, motor vehicles, and other sources

- Road salts.

These pollutants can harm fish and wildlife populations, kill native vegetation, foul drinking water supplies, and make recreational areas unsafe and unpleasant.

Stream Damage

Increased flow resulting from excess runoff causes extensive damage to streams, scouring the stream bottom and causing severe erosion to stream banks. The risk of flooding downstream also increases.

Stream health

Urban Runoff

Urban runoff is either wet weather (rainwater) or dry weather (water waste) flows from urban landscapes into storm drain systems that lead to the beach.

Urban Runoff: A Problem

Urban runoff carries contaminants, such as litter, food, human & animal waste, automobile fluids, industrial pollutants, fertilizers and pesticides to the beach creating health risks for people, killing marine life and contributing to localized flooding and beach closures.

We also see the impacts of pollution in increased health risks to swimmers near flowing storm drains and toxicity to aquatic life. These impacts translate into losses to the tourism economy, loss of recreational resources, dramatic cost increases for cleaning up contaminated sediments, degraded water quality and impaired function and vitality of our natural resources.

Clogged storm drains can lead to area flooding when it rains, creating traffic problems and unsanitary conditions.

Urban Flooding

Flooded streets in New Orleans

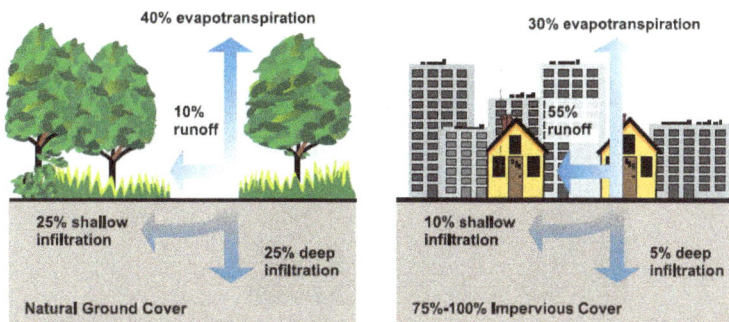

Relationship between impervious surfaces and surface runoff

Urban runoff is a major cause of urban flooding, the inundation of land or property in a built-up environment caused by rainfall overwhelming the capacity of drainage systems, such as storm sewers. Triggered by events such as flash flooding, storm surges, overbank flooding, or snow melt, urban flooding is characterized by its repetitive, costly and systemic impacts on communities, regardless of whether or not these communities are located within formally designated floodplains or near any body of water.

There are several ways in which stormwater enters properties: backup through sewer pipes, toilets and sinks into buildings; seepage through building walls and floors; the accumulation of water on property and in public rights-of-way; and the overflow of water from water bodies such as rivers and lakes. Where properties are built with basements, urban flooding is the primary cause of basement flooding.

Flood flows in urban environments have been investigated relatively recently despite many centuries of flood events. Some researchers mentioned the storage effect in urban areas. Several studies looked into the flow patterns and redistribution in streets during storm events and the implication in terms of flood modelling. Some recent research considered the criteria for safe evacuation of individuals in flooded areas. But some recent field measurements during the 2010–2011 Queensland floods showed that any criterion solely based upon the flow velocity, water depth or specific momentum cannot account for the hazards caused by the velocity and water depth fluctuations. These considerations ignore further the risks associated with large debris entrained by the flow motion.

Effects

A 2008 report by the United States National Research Council identified urban runoff as a leading source of water quality problems.

> ...further declines in water quality remain likely if the land-use changes that typify more diffuse sources of pollution are not addressed. These include land-disturbing agricultural, silvicultural, urban, industrial, and construction activities from which hard-to-monitor pollutants emerge during wet-weather events. Pollution from these landscapes has been almost universally acknowledged as the most pressing challenge to the restoration of waterbodies and aquatic ecosystems nationwide.
>
> – National Research Council, *Urban Stormwater Management in the United States*

Weasel Brook in Passaic, New Jersey has been channelized with concrete walls to control localized flooding.

The runoff also increases temperatures in streams, harming fish and other organisms. (A sudden burst of runoff from a rainstorm can cause a fish-killing shock of hot water.) Also, road salt used to melt snow on sidewalks and roadways can contaminate streams and groundwater aquifers.

One of the most pronounced effects of urban runoff is on watercourses that historically contained little or no water during dry weather periods (often called *ephemeral streams*). When an area around such a stream is urbanized, the resultant runoff creates an unnatural year-round stream-flow that hurts the vegetation, wildlife and stream bed of the waterway. Containing little or no sediment relative to the historic ratio of sediment to water, urban runoff rushes down the stream channel, ruining natural features such as meanders and sandbars, and creates severe erosion—increasing sediment loads at the mouth while severely incising the stream bed upstream. As an example, on many Southern California beaches at the mouth of a waterway, urban runoff carries trash, pollutants, excessive silt, and other wastes, and can pose moderate to severe health hazards.

Because of fertilizer and organic waste that urban runoff often carries, eutrophication often occurs in waterways affected by this type of runoff. After heavy rains, organic matter in the waterway is relatively high compared with natural levels, spurring growth of algae blooms that soon consume most of the oxygen. Once the naturally occurring oxygen in the water is depleted, the algae blooms die, and their decomposition causes further eutrophication. Algae blooms mostly occur in areas with still water, such as stream pools and the pools behind dams, weirs, and some drop structures. Eutrophication usually comes with deadly consequences for fish and other aquatic organisms.

Excessive stream bank erosion may cause flooding and property damage. For many years governments have often responded to urban stream erosion problems by modifying the streams through construction of hardened embankments and similar control structures using concrete and masonry materials. Use of these hard materials destroys habitat for fish and other animals. Such a project may stabilize the immediate area where flood damage occurred, but often it simply shifts the problem to an upstream or downstream segment of the stream.

Urban flooding has significant economic implications. In the US, industry experts estimate that wet basements can lower property values by 10%-25% and are cited among the top reasons for not purchasing a home. According to the U.S Federal Emergency Management Agency (FEMA), almost 40% of small businesses never reopen their doors following a flooding disaster. In the UK, urban flooding is estimated to cost £270 million a year in England and Wales; 80,000 homes are at risk.

A study of Cook County, Illinois, identified 177,000 property damage insurance claims made across 96% of the county's ZIP codes over a five-year period from 2007-2011. This is the equivalent of one in six properties in the County making a claim. Average payouts per claim were $3,733 across all types of claims, with total claims amounting to $660 million over the five years examined.

Despite concerted efforts, many communities lack the funds to fully address these issues, and often seek funds elsewhere. Numerous watersheds within Los Angeles County, California do not meet state water quality standards, despite spending $100 million a year on clean water programs to combat issues such as urban runoff. To combat this problem, officials have introduced a measure that would assess a fee to homeowners and local businesses in attempt to raise $290 million for effective urban runoff management.

Prevention and mitigation

Effective control of urban runoff involves reducing the velocity and flow of stormwater, as well as reducing pollutant discharges. A variety of stormwater management practices and systems may be used to reduce the effects of urban runoff. Some of these techniques (called best management practices (BMPs) in the US), focus on water quantity control, while others focus on improving water quality, and some perform both functions.

A percolation trench infiltrates stormwater through permeable soils into the groundwater aquifer.

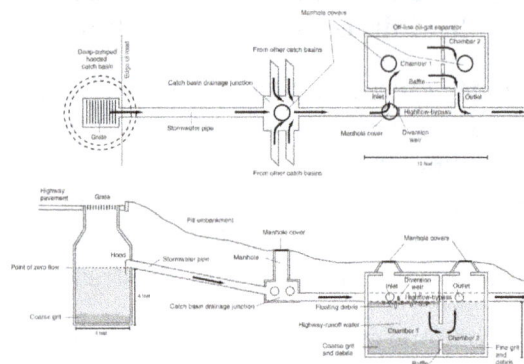

An oil-grit separator is designed to capture settleable solids, oil and grease, debris and floatables in runoff from roads and parking lots

Pollution prevention practices include low impact development (LID) or green infrastructure techniques - known as Sustainable Drainage Systems (SuDS) in the UK, and Water-Sensitive Urban Design (WSUD) in Australia and the Middle East - such as the installation of green roofs and improved chemical handling (e.g. management of motor fuels & oil, fertilizers and pesticides). Runoff mitigation systems include infiltration basins, bioretention systems, constructed wetlands, retention basins and similar devices.

Providing effective urban runoff solutions often requires proper city programs that take into account the needs and differences of the community. Factors such as a city's mean temperature, precipitation levels, geographical location, and airborne pollutant levels can all effect rates of pollution in urban runoff and present unique challenges for management. Human factors such as urbanization rates, land use trends, and chosen building materials for impervious surfaces often exacerbate these issues.

The implementation of citywide maintenance strategies such as street sweeping programs can also be an effective method in improving the quality of urban runoff. Street sweeping vacuums collect particles of dust and suspended solids often found in public parking lots and roads that often end up in runoff.

Educational programs can also be an effective tool for managing urban runoff. Local businesses and individuals can have an integral role in reducing pollution in urban runoff simply through their practices, but often are unaware of regulations. Creating a productive discussion on urban runoff and the importance of effective disposal of household items can help to encourage environmentally friendly practices at a reduced cost to the city and local economy.

Blue drain and yellow fish symbol used by the UK Environment Agency to raise awareness of the ecological impacts of contaminating surface drainage

Runoff Model

Outflow at the bottom of the soilwater box, f, forms inflow to the runoff box, i.e it corresponds to peff, as it has been described in the section on reservoir models. A conceptual runoff model can therefore be illustrated as in figure below. One has to solve the continuity equations for soilwater eq. and for the runoff box eq. ($\frac{ds}{dt} = p - e - q$) and the accompanying runoff (the equations are given new numbering here):

$$\frac{d_{h_{mark}}}{d\,t}2 = p - e - f; e, f = funk\left(h_{mark}\right)$$

$$\frac{dh}{d\,t}\,3 = f\,-\,q;\,q\,=\,funk\left(h\right)$$

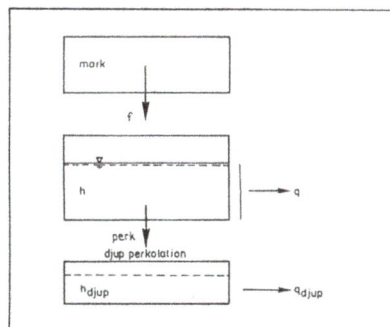

Figure: Runoff model with soilwater reservoir and double runoff reservoir

The solution is most easily done with an explicit method according to the scheme given in figure above. hmark is related to hwilt, which is taken to be zero.

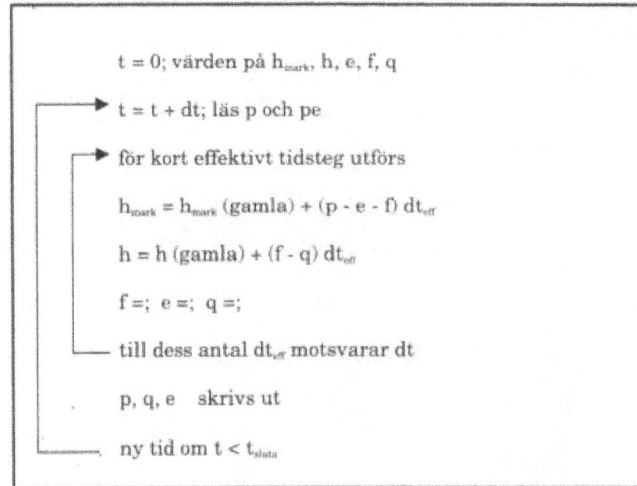

Figure: Explicit solution in runoff model

The runoff model can be made more complex than shown above. In the model shown in figure above there are 5 parameters, hp; FC; To; T1; ho. The more parameters introduced the more difficult it is to adjust the model to real general situations even if special events can be modeled almost perfectly. One can include surface runoff from the soilwater box, one can use several holes, i.e. more time constants, in the runoff box. One can allow evaporation directly from parts of the runoff box and one can let some precipitation go straight to the runoff box. One can let runoff go into another box. There are several modifications all of which can be motivated in special situations.

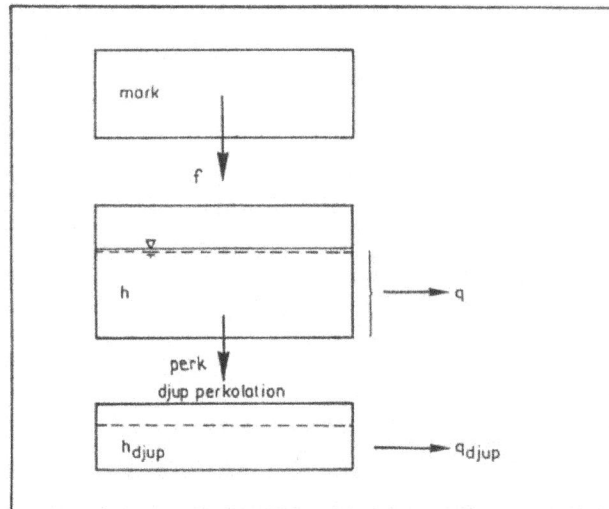

Figure: Runoff model with soilwater box , runoff reservoir and a deep water storage

An addition to the runoff model shown in figure above which is very often used is a deep groundwater box, which is used to model slow groundwater flow forming base flow. Water is supposed to percolate from the runoff box into the deep groundwater box, as shown in figure above. In the Swedish HBV-model smaller lakes are taken as part of the groundwater storage in the sense that precipitation on the lake surface goes directly to deep storage and evaporation from the lake surface constitutes a loss from the deep storage. The rate of percolation, perk, from groundwater box to deep groundwater-box can be constant or depend on amount of stored water. The continuity equations for the runoff box and the deep groundwater box becomes respectively:

$$\frac{dh}{dt}4 = f - q - perk$$

$$\frac{d\,h_{djup}}{dt}5 = perk - q_{djup}$$

$$qd_{djup} = \frac{h_{djup}}{T_{djup}}6$$

The runoff is the summation of q and d_{djup}

Before one has a complete runoff model the three boxes described here for soilwater, groundwater and deep groundwater must be supplemented with interception- and snow models for calculation of inflow to the soilwater box, and with a routing model which describes how the runoff is delayed in the system of streams and rivers in the catchment.

Runoff Curve Number Method

The SCS Runoff Curve Number method is developed by the United States Department of Agriculture (USDA) Soil Conservation Service (SCS) and is a method of estimating rainfall excess from rainfall.

The conceptual basis of the curve number method has been the object of both support and criticism (Ponce and Hawkins, 1996). The major disadvantages of the method are sensitivity of the method to Curve Number (CN) values, fixing the initial abstraction ratio, and lack of clear guidance on how to vary Antecedent Moisture Conditions (AMC). However, the method is used widely and is accepted in numerous hydrologic studies. The SCS method originally was developed for agricultural watersheds in the mid-western United States; however it has been used throughout the world far beyond its original developers would have imagined.

The basis of the curve number method is the empirical relationship between the retention (rainfall not converted into runoff) and runoff properties of the watershed and the rainfall. Mockus found equation 1 appropriate to describe the curves of the field measured runoff and rainfall values (National Engineering Handbook, 2004). Equation 1 describes the conditions in which no initial abstraction occurs.

$$\frac{F}{S} = \frac{Q}{P}$$

where F = P − Q = actual retention after runoff begins;

Q = actual runoff

S = potential maximum retention after runoff begins (S ³ F)

P = potential maximum runoff (i.e., total rainfall if no initial abstraction).

For most applications, a certain amount of rainfall is abstracted. The three important abstractions for any single storm event are rainfall interception (Meteorological rainfall minus throughfall, stem flow and water drip), depression storage (topographic undulations), and infiltration into the soil. The curve number method lumps all three abstractions into one term, the Initial abstraction (Ia), and subtracts this calculated value from the rainfall total volume. The total rainfall must exceed this initial abstraction before any runoff is generated. This gives the potential maximum runoff (rainfall available for runoff) as P − Ia. Substituting this value in equation 1 yields following equation:

$$\frac{P - Ia - Q}{S} = \frac{Q}{P - Ia}$$

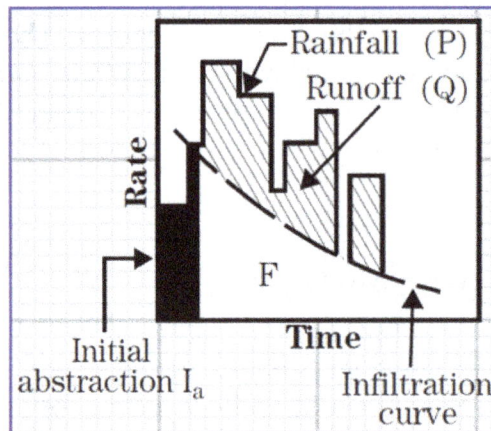

Figure: Components of SCS Runoff equation

It is important to note the potential maximum retention term, "S", excludes Ia. Hence, for a given storm, maximum loss of rainfall is S plus Ia. Rearranging terms in Equation ($\frac{P - Ia - Q}{S} = \frac{Q}{P - Ia}$) for Q gives

$$Q = \frac{(P - Ia)^2}{(P - Ia) + S}$$

Establishing the relation to estimate Ia was challenging. The SCS provided the following empirical equation below based on the assumption Ia was a function of the potential maximum retention S.

Ia = 0.2S

The potential maximum retention S is related to the dimensionless parameter CN in the range of 0 <= CN <= 100 by equation below.

$$S = \left(\frac{1000}{CN}\right) - 10)$$

Substituting Equation (Ia = 0.2S) into Equation ($Q = \dfrac{(P-Ia)^2}{(P-Ia)+S}$) yields,

$$Q = \frac{(P-0.2S)^2}{(P+0.8S)}$$

Equation above has only one parameter that needs to be evaluated (i.e., S) which can be determined by using equation ($S = \left(\dfrac{1000}{CN}\right)-10$) and curve number tables published by the SCS.

The solution of the SCS runoff equation is shown below.

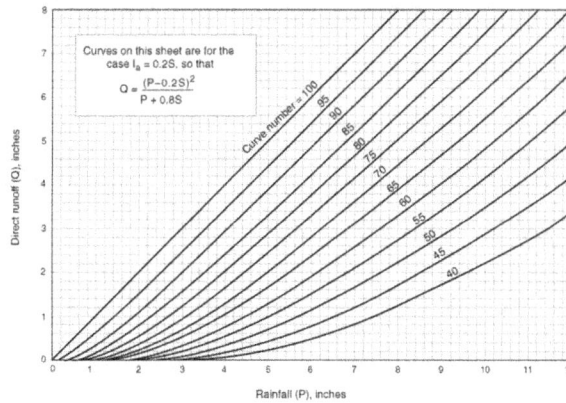

When the drainage area consists of land use with impervious cover (Directly Connected or Unconnected), TR 55 provides separate graphs for computing the composite curve number values depending on the percent of the impervious cover. However, it is a good practice to calculate the runoff from impervious area and pervious area separately and then add the volume.

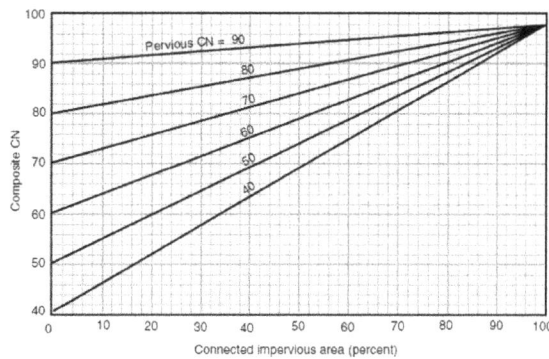

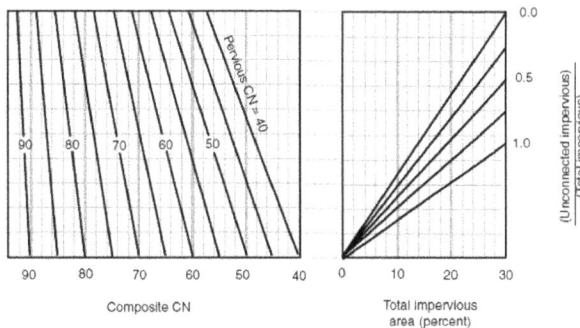

References

- National Research Council (United States) (2009). Urban Stormwater Management in the United States (Report). Washington, D.C.: National Academies Press. p. 24. doi:10.17226/12465. ISBN 978-0-309-12539-0.

- Fundamentals: physicalgeography.net, Retrieved 28 June 2018

- Ballo, Siaka; Liu, Min; Hou, Lijun; Chang, Jing (2009-07-10). "Pollutants in stormwater runoff in Shanghai (China): Implications for management of urban runoff pollution". Progress in Natural Science. 19 (7): 873–880. doi:10.1016/j.pnsc.2008.07.021.

- What-is-snowmelt: albertawater.com, Retrieved 29 May 2018

- "Parking Lot and Street Cleaning". National Menu of Stormwater Best Management Practices. EPA. Archived from the original on 2015-08-28. Retrieved 2014-12-24.

- Watercyclesnowmelt: water.usgs.gov, Retrieved 30 March 2018

- Sansalone, John; Christina, Chad M. (November 2004). "First flush concepts for dissolved solids in small impervious watersheds". Journal of Environmental Engineering. 130 (11): 1301–1314. doi:10.1061/(ASCE)0733-9372(2004)130:11(1301). ISSN 0733-9372.

- Stormwater-education: villagehohny.org, Retrieved 10 May 2018

- G. Allen Burton, Jr., Robert Pitt (2001). Stormwater Effects Handbook: A Toolbox for Watershed Managers, Scientists, and Engineers. New York: CRC/Lewis Publishers. ISBN 0-87371-924-7.Chapter 2.

- Stormwater: rivanna-stormwater.org, Retrieved 22 June 2018

- Werner, MGF; Hunter, NM; Bates, PD (2006). "Identifiability of Distributed Floodplain Roughness Values in Flood Extent Estimation". Journal of Hydrology. 314: 139–157. doi:10.1016/j.jhydrol.2005.03.012.

- Curve-number-introduction: professorpatel.com, Retrieved 30 March 2018

- Schueler, Thomas R. (2000) [initial publ. 1995]. "The Importance of Imperviousness" (pdf). In Schueler; Holland, Heather K. The Practice of Watershed Protection. Ellicott City, MD: Center for Watershed Protection. pp. 1–12. Retrieved 2014-12-24.

Infiltration Hydrology

The process by which any surface water seeps into the soil is known as infiltration hydrology. It is caused due to the gravity and capillary action. Infiltration rate and capacity is affected by various factors such as vegetation cover, soil texture, soil temperature, rainfall intensity, etc. This is an important chapter, which analyzes infiltration hydrology, field capacity, soil plant atmosphere continuum, etc.

Infiltration

Infiltration is the process by which precipitation or water soaks into subsurface soils and moves into rocks through cracks and pore spaces. The bulk of rainwater and melted snow end up infiltrated.

Infiltration Characteristics

The infiltration capacity is the maximum rate at which water can be absorbed by a given soil per unit area under given conditions.

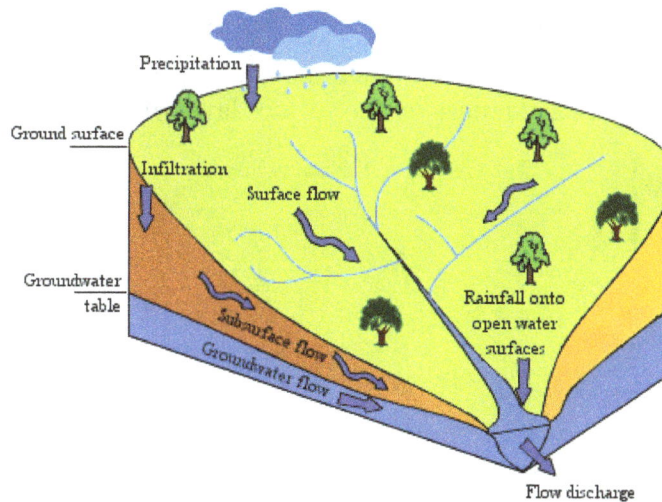

Figure: The infiltration process depending on soil type and flow

$$I(t) = \int_{t-t_d}^{t} i(t)dt$$

where:

> *I(t)* the cumulative infiltration during the t period (mm)
>
> *i(t)* the infiltration regime during the t period (mm/h)

Hydraulic conductivity at saturation k_s, is an essential parameter of infiltration. It represents the limiting value of infiltration if the soil is saturated and homogenous. Percolation is the vertical water flow in soils (porous unsaturated environment) on the groundwater layer under the influence of gravity. This process follows infiltration and has a major influence on the underground layer water supply.

Net rain is the amount of rain that falls to the ground surface during a shower. The clear rain is deduced from the total rain, diminished by the intercepted fraction of vegetation and that which is stored in ground depressions. The difference between the infiltrated rain and the drained rain on the ground surface is called production function.

Factors which influence infiltration

The main factors that influence the infiltration are:

- The soil type (texture, structure, hydrodynamic characteristics). The soil characteristics influence capillary forces and adsorption;

- The soil coverage. Vegetation has positive influence on infiltration by increasing the time of water penetration in soil;

- The topography and morphology of slopes;

- The flow supply (rain intensity, irrigation flow);

- The initial condition of soil humidity. Soil humidity is an important factor of infiltration regime. The infiltration regime evolves differently in time for dry or wet soils;

- Soil compaction due to rain drop impact and other effects. The use of hard agricultural equipment can have consequences on the surface layer of soil.

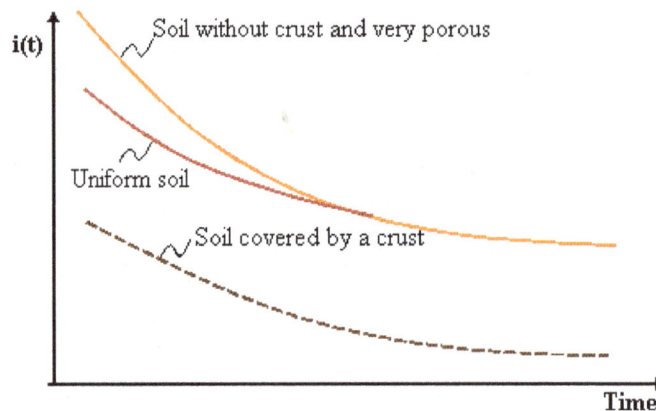

Figure: The infiltration regime depending on time for different types of soil

Models Used to Estimate Infiltration Rates

Infiltration processes can be estimated by means of different models:

- Models based on empirical relations involving 2, 3 or 4 parameters;

- Physically based models.

Models based on Empirical Relations

Empirical relations show a decrease of infiltration depending on initial time (either exponential or quadratic function of time), which tends to a limit value, generally k_s, but near o. An empirical relation is the Horton formula, where the infiltration capacity can be expressed as following:

$$I(t) = i_f + \left(i_0 - i_f\right)e^{-\gamma t}$$

$$I(t) = \int_0^t i(t)dt = i_0 t + \frac{i_0 - i_f}{\gamma}\left(1 - e^{-\gamma t}\right)$$

where:

- $i(t)$ infiltration capacity at time t (mm/h)
- i_o initial infiltration capacity depending on soil type (mm/h)
- t time from the beginning of the shower (h)
- $I(t)$ total quantity of infiltrated water from initial time until the moment t (mm water column)
- γ empirical constant depending on soil type (min⁻¹)

This formula is not linear and it presents certain practical difficulties. Through linearization, we obtain:

$$\frac{i(t) - i_f}{i_0 - i_f} = e^{-\gamma t}$$

As logarith, we get:

$$In\frac{i(t) - i_f}{i_0 - i_f} = -\gamma t$$

The formula of the Institute of Soil and Water Management of the EPFL is:

$$i(t) = i_f + ae^{-\gamma t}$$

where:

- $i(t)$ infiltration capacity at time t (mm/h)
- i_f final infiltration capacity (mm/h)
- a,b   correction coefficients

The relation is a little different from that of Horton. There are just two parameters. This relation has the advantage of allowing the search of functional relations between the limit/final capacity of infiltration and soil texture, and also between the parameter a and the amount of soil humidity. Other formulas can be used to determine the infiltration regime of water from soil.

Physically based Models

These models describe in a simplified manner the water movement in soils, especially at the humidity front level, depending on certain physical parameters.

Table:　Main functions used at infiltration

Author	Function	Legend
Horton	$i(t) = i_f + (i_0 - i_f)e^{-\gamma}$	$i(t)$ - infiltration capacity during time [cm/s] i_o - initial infiltration capacity [cm/s] i_f - final infiltration capacity [cm/s] γ - constant depending on the soil type
Kostiakov	$i(t) = i_0 t^{-\alpha}$	α - parameter depending on soil conditions
Dvorak-Mezencev	$i(t) = i_0 + (i_1 - i_f)t^{-b}$	i_1 - infiltration capacity at time t=1min [cm/s] t - time [s] b - constant
Holtan	$i(t) = i_f + cw[(IMD) - F]^n$	c - factor variable from 0.25 to 0.8 w - Holtan equation flow factor n - experimental constant approximately = 1.4
Philip	$i(t) = \dfrac{1}{2}st^{-0.5} + A$	s - sorptivity [cms$^{-0.5}$] A - gravity component depending on hydraulic conductivity at saturation [cm/s]
Dooge	$i(t) = a(F_{max} - F_t)$	a - constant F_{max} - maximal retention capacity F_t - water quantity retained on soil at time t
Green & Ampt	$i(t) = k_s \left(1 + \dfrac{h_0 - h_f}{z_f(t)}\right)$	k_s - hydraulic conductivity at saturation [mm/h] h_o - surface pressure load [mm] h_f - pressure load at the humidity front [mm] z_f - humidity front depths [mm]

From the models presented in table above the following two models have been used most often:

The Philip Model

Philip proposed a method of resolving the vertical infiltration for certain initial and boundary conditions. This model has introduced the notion of "sorption" that represents the soil capacity to absorb water when the flow is produced only under gradient pressure. The infiltration can be simplified as follows:

$$i(t) = \frac{1}{2}st^{-0.5} + A$$

where:

$i(t)$　infiltration rate (cm s^{-1})

s　sorption (cm s$^{-0.5}$)

t time (s)

A the gravity component, depending on hydraulic conductivity on saturation (cm s^{-1})

For horizontal infiltration the gravity gradient is not involved. Infiltration will have the following expression:

$$i(t) = \frac{1}{2} s t^{-0.5}$$

$$s = \frac{I}{\sqrt{t}}$$

The Green and Ampt Model

This model is based on hypotheses that involve a schematisation of infiltration processes.

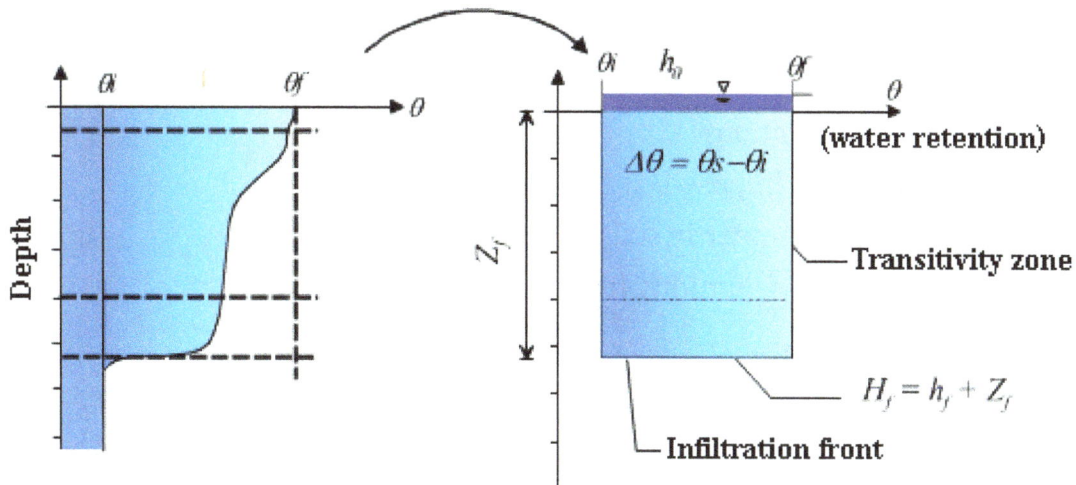

Figure: Infiltration process schematisation according to Green and Ampt

The method's main hypotheses are:

- A humidity front is perfectly defined;

- A transmission zone, where in time and space water storage and hydraulic conductivity are constant;

- The suction forces of the humidity front are constant;

Based on the Darcy law the model includes the hydrodynamic parameters of soil:

$$i(t) = \frac{\partial I}{\partial t} = k_s \frac{H_0 - H_f(t)}{z_f(t)} = k_5 \frac{h_0 - h_f z_f(t)}{z_f(t)}$$

where:

$i(t)$ infiltration rate (mm)

t time (h)

k_s hydraulic conductivity at saturation (mm/h)

H_o hydraulic total head at surface (mm)

$H_f(t)$ hydraulic total head at the humidity front level (mm)

z_f maximum depth of the humidity front

h_o pressure head at surface (mm)

h_f pressure head of the humidity front (mm)

One of Green's and Ampt's model hypotheses stipulates that water storage from the transition zone is uniform. The cumulative infiltration $I(t)$ results from the product of water storage and the depth of the humidity front.

$$i(t) = -(\theta_0 - \theta) \cdot z_f(t)$$

where:

$I(t)$ cumulate infiltration

θ_o quantity of water imposed on the surface

θ initial quantity of water in soil

z_f maximum depth of humidity front

The last two relations result in:

$$i(t) = \frac{\partial I(t)}{\partial t} = -(\theta_0 - \theta_i)\frac{\partial z_f(t)}{\partial t} = k_s \frac{h_0 - h_f - z_f(t)}{z_f(t)}$$

For horizontal flow infiltration has the relation:

$$i(t) = k_s \frac{(\theta_0 - \theta_f)(h_0 - h_f)}{I(t)}$$

For vertical flow infiltration becomes:

$$i(t) \approx -k_s$$

This model is satisfactory when applied to a soil with coarse texture, but it is an empirical method in which it is necessary to determine the pressure head of humidification front.

Variation of Infiltration Rates During a Rainfall

The spatial and temporal variability of water quantity existing in soil is described by infiltration curves or hydric profile. These represent the vertical water distribution in soil at different periods t. In a homogeneous (uniform) soil when the soil surface is flooded, the hydric profile has three zones: a saturation zone, a transition zone and a humidity zone.

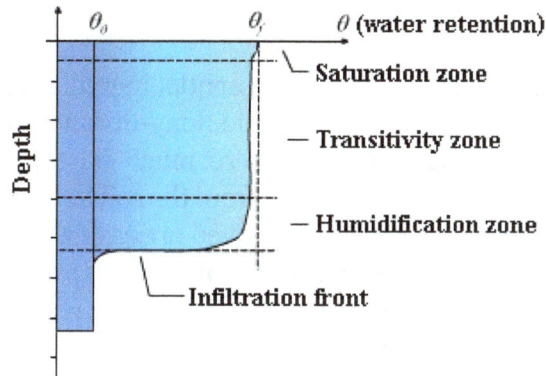

Figure: Characteristics of the hydric profile during infiltration

During a rainfall the infiltration capacity of soil decreases to a limiting value, which represents the infiltration potential at saturation. If we compare the rain intensity and the infiltration capacity of the soil, there are two possibilities:

- When the rain intensity is inferior to infiltration capacity, water infiltrates faster due to the supply regime. The necessary time to equalize the infiltration capacity is variable and depends on existing soil humidity conditions or on the shower. The time taken is longer when the soil is dry and the water supply regime is similar to the hydraulic conductivity at saturation k_s;

- When the rain intensity is superior to the infiltration capacity of the soil the water surplus is stocked on the surface or in ground depressions. The infiltration regime and the infiltration capacity for net storm rain are presented in the next Figure.

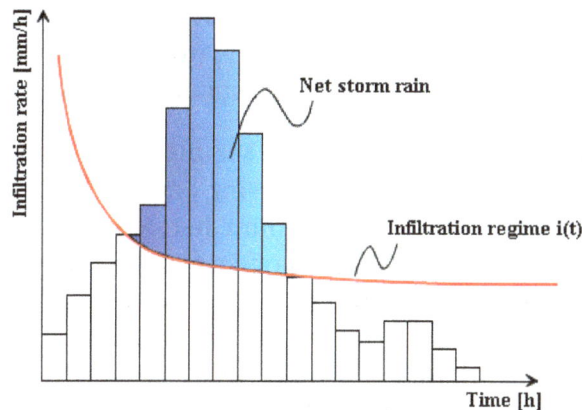

Figure: Infiltration regime and net storm rain

Soil Plant Atmosphere Continuum

The dominant distinguishing feature of the Earth, in comparison with all other celestial bodies we observe, is the presence of life. Humans, as a part of life on Earth, attempt to understand all of the life system, albeit from an anthropocentric perspective.

As our integrative understanding of living systems within the soil–plant–atmosphere continuum (SPAC) grows, the idea that a whole ecosystem can function as a living entity becomes more acceptable.

Consider a terrestrial ecosystem on land: The most essential ingredient for life on Earth is water and most of the water is not on the land. However, a rapid, long-distance system for transporting water from oceans to land is available through the atmosphere, much like the xylem system in a tree, which transports water long distances (relative to cell sizes) from the soil to the leaves high above the ground. This long-distance atmospheric transport system is driven unceasingly by the uneven, solar heating of the Earth. On the land, where many organisms reside, water moves much more slowly through the soil than through the atmosphere and this provides an essential storage reservoir that maintains the critical continuous flow of water to plants in the presence of fluctuations in the atmospheric supply. The rapid atmospheric transport system that is so critical to moving water from the ocean to the land also fluctuates wildly over the life span of most organisms. The combination of a low-storage, highly fluctuating system for rapid global transport in the atmosphere and a large-storage, stable, low-flow buffer in the soil redistributes water over the Earth to sustain life on land. This is a highly structured system like an organism. Furthermore, the living systems exert a strong influence on this larger system to enhance their functioning, just as we humans modify our environment to improve our lot.

Water is the most obvious component of the SPAC, and the interaction of this water with energy is the cornerstone of our understanding of how life interacts with its environment. Every schoolchild learns the water cycle: evaporation of liquid water from surfaces supplies water vapor to the atmosphere, where air currents lift this water to heights that cool it and cause condensation of the water vapor back to the liquid form of droplets, which fall back to the Earth to be recycled. The extraction of energy from the surface as a result of evaporation and the release of energy to the atmosphere by condensation represent the primary means by which the energy from the sun ultimately warms the atmosphere. This 'latent heat,' which is associated with the change of phase of water between liquid and vapor states, and not a temperature change, provides the key feedback between water and energy cycles that stabilizes the soil–plant–atmosphere system, permitting it to sustain life.

Materials such as water flow through the ecosystem seamlessly, moving from place to place, serving innumerable functions, and consuming or releasing energy to achieve a phase suitable to the medium that contains it. This is the SPAC; 'continuum' here does not refer to the material storage forms so apparent to our eyes such as plant leaves or roots, lakes, raindrops, the soil itself, or even the air we breathe; rather it refers to the unimpeded flow of many forms of material and energy throughout an elegant system Soil–plant–atmosphere continuum 513 for sustaining life on land. Animals are as much a part of this system as plants. This ceaseless flow of material and energy sustains life at every level; from the molecular interactions that sustain and mutate the genetic 'memory,' to cellular processes that make up whole organisms, to the global level of a living, breathing planet.

Living systems cycle matter through the SPAC and themselves to create and maintain their structure and extract energy for powering their life functions. All material necessary for sustaining life is cycled on all spatial levels from subcellular to global. The connectedness and functioning of the SPAC can be illustrated by reference to water, the most critical mass constituent to life on Earth. However, understanding the SPAC through water will necessarily involve some considerations of at least carbon and energy.

Fundamental Principles

Two general principles are indispensable for studying the SPAC: (1) the conservation principles that take the form of mass and energy 'budgets,' and (2) the transport principles that relate the flow of some quantity to the difference or gradient of other quantities that influence or 'force' the flow and describe the 'state' of the exchange process.

The principles of conservation of mass and energy are the backbone of integrative studies of the SPAC. Because mass and energy can take many forms as they move throughout the SPAC, and even interact with each other, budgets are constructed to quantify the important stores and flows of important life-enabling constituents such as water, carbon, or energy. A budget is simply the application of the conservation principles (mass or energy) to a specific system that must be carefully defined. This system may be defined as a leaf, a community of plants, or even the entire Earth. Like balancing a bank account, a budget is a formal statement of the following: Incoming quantity minus outgoing quantity equals the increase in storage of the quantity in the system. In the case of water (W), carbon (C), and energy (E), the budgets become the following:

$$W_{IN} - W_{OUT} = \Delta W$$

$$C_{IN} - C_{OUT} = \Delta C$$

$$E_{IN} - E_{OUT} = \Delta E$$

The intertwined budgets of water, carbon and energy inflow, outflow and storage within the soil–plant–atmosphere system are shown in figure below

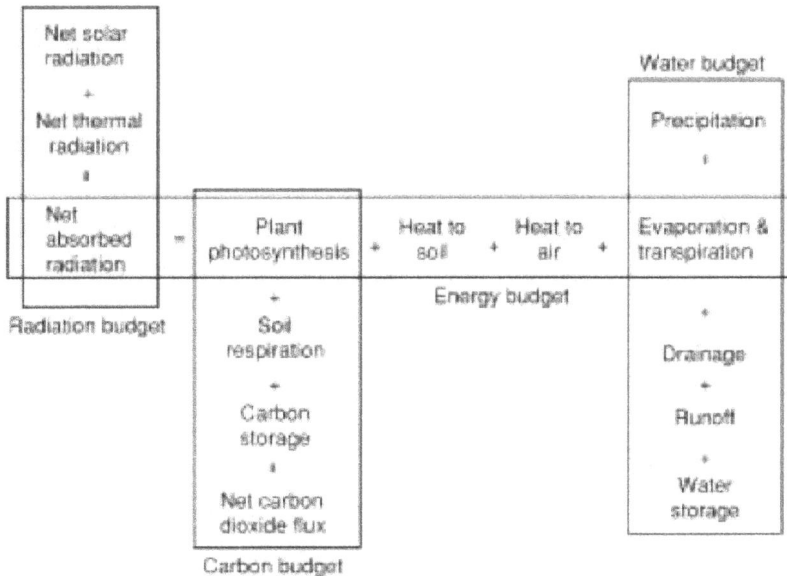

Figure: The interlocking energy and mass budgets for water, carbon, and energy in the soil–plant–atmosphere continuum. Altering individual components can cascade through the system and affect many seemingly unrelated factors.

Understanding the SPAC through the role of water will involve simultaneous consideration of carbon and energy budgets on a spatial scale appropriate for living systems.

Consider the energy budget: Energy can exist in many forms as it flows through the soil–plant–atmosphere system, and the energy budget represents a way to keep track of the energy, no matter what its form. The rate of energy flow (joules per second ¼ watts) per unit of area (meters squared) is a convenient way to describe the energy budget for a particular land area, because the size of the area does not matter and the emphasis can be on the energy exchanges. The radiant energy from the sun and surroundings can be absorbed by a leaf (this is referred to as the net radiation) and be transformed to the following: (1) sensible heat energy that is convected away from the leaf by moving air that is cooler than the leaf; (2) latent energy in molecules of water that are converted from liquid to vapor inside the leaf as they are transpired (evaporated) from the leaf; and (3) biochemical energy in the form of organic compounds such as sugars and starches created through photosynthesis and other physiological processes to sustain life. These components of the leaf energy budget are shown in figure above.

The transport principle simply states that the flow of some quantity such as water is equal to an appropriate 'driving force' divided by a resistance to transport exerted by the various media through which the quantity moves. The transport principle is more difficult to use than the conservation principle because the 'driving forces' and transport resistances depend on the characteristics of the system and mechanism of transport. For example, the diffusion of liquid depends on the water potential difference across a hydraulic transport resistance, whereas the transport of water vapor depends on the water vapor pressure difference across a diffusive transport resistance. In addition, movement of liquids or vapor by mass flow is 'driven' by pressure differences across transport resistances. In general, diffusion is an effective mode of transport over small distances, but mass flow can move materials and energy over longer distances. Heat transport is driven by temperature differences. The relevant thermal resistance depends on whether the heat is moving through a solid such as soil (conduction, which is analogous to diffusion), or through a moving fluid such as air (convection, which is analogous to mass flow and is a much more rapid transport mechanism than diffusion). Heat transport by radiation depends on still-different formulations. Thus much research on the SPAC relates to determining appropriate transport resistances and state variables for characterizing the flow of important quantities throughout the system. A complete understanding of this soil–plant–atmosphere system is a formidable task, but such an understanding will improve the ability of humans to make choices that will sustain a diverse and resilient ecosystem, upon which our survival depends. The fundamental concepts of energy and mass conservation and transport relations constitute the theoretical backbone of environmental biophysics (the study of the SPAC).

Water in the Soil–Plant–Atmosphere-Continuum

The smallest spatial scale that is usually associated with the SPAC is a field plot (square meters to thousands of square meters) from the bottom of a root zone (a few meters deep) to a few tens of meters above the top of the vegetation. Although individual studies may use smaller systems such as individual leaves or potted plants to reveal mechanisms, the scale of at least a small field is required to encompass most of the essential components of the natural SPAC and yet be accessible to direct measurement. Researchers generally use formulations nearly identical to those measured in field plots to represent idealized patches of the Earth's surface thousands of square kilometers in size; then they combine these patches to understand better the human influence on global climate processes. Typical time scales for small-scale studies vary from minutes to months or even years. On these spatial and temporal scales, researchers attempt to measure all the important flow and state variables, which are used to validate models that can be applied in systems that cannot be measured directly. After all, society cannot afford to measure everything everywhere, but will support measuring some things in many places and many things in a few places.

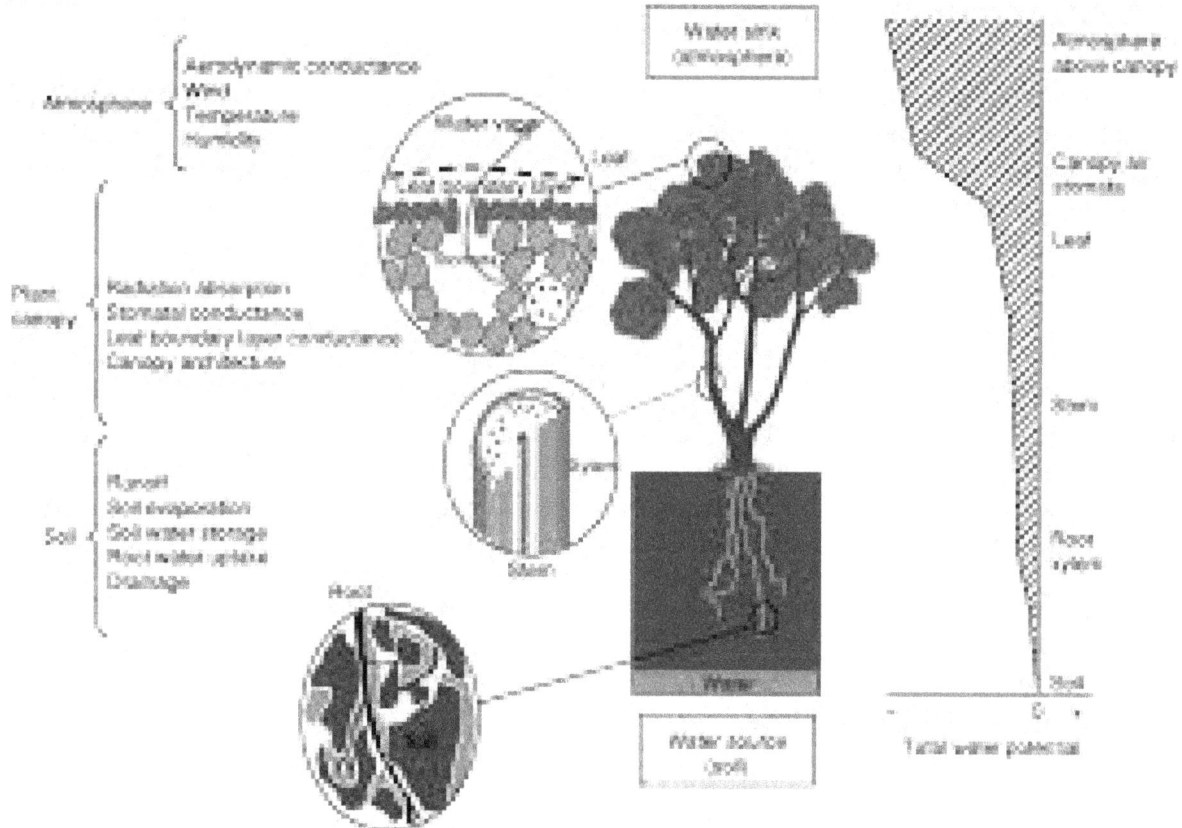

Figure: Water movement and storage in the soil–plant–atmosphere continuum. The graph on the right indicates the decrease in water potential that occurs as water moves from the soil to the atmosphere. On the left the three components of the soil–plant–atmosphere system are listed with the major factors for each that should be considered in a model of intermediate complexity.

The transport and storage of water in the SPAC are shown in figure above. The flow of liquid water from the soil into the plant depends on the plant having a lower water potential than the soil; thus the soil water potential determines the availability of soil water for diffusion into the plant. The water potential is the amount of energy (joules) that can be obtained from moving a mass of water (kilograms) to a pool of pure water at atmospheric pressure and specified elevation; and in the soil and plant it is usually negative, indicating that energy must be expended to remove the water. As soils dry, the water potential decreases, making it more difficult for plants to obtain the water they need to grow. The soil water potential is related to the water content of the soil through the 'moisture-release curve,' which varies from soil to soil. Moisture-release curves have been measured for many different soil types. Typically soils can provide an amount of water to plants from storage that is equivalent to 5% (sands) to 20% (silt loams) of the plants' rooting depth. Thus for a plant rooted to a depth of 1 m, soils can provide approximately 50–200 mm of water to the plant. With evapotranspiration rates of 2–10 mm day1, soils can store enough water to sustain plants for a month or more without rainfall or irrigation.

Once water enters the plant through its roots, it moves through the plant vascular system to the leaves if the leaves have lower water potentials than the roots. Since the transport resistance to mass flow in the stem xylem is usually less than the transport resistance to diffusion through the root, and the flows through both root and xylem must be equal according to conservation principles, the water potential difference from the soil to the root xylem will usually be greater than the water potential difference from the root xylem to the leaf.

Water is lost from plants through pores in the leaf surfaces, called stomata. Evaporation from cells just below the stomatal pores draws water from the leaf cells, which in turn are hydraulically connected to a continuous water column down to the roots. As water is transpired through the stomata, additional water is drawn through the xylem to replace it, very much like suction applied to a straw in a soda. Of course, this evaporation that occurs just below the leaf surface consumes most of the Sun's energy that is absorbed by the leaf, because water vapor molecules carry away latent energy that was not contained in the liquid water that came up from the roots. While the flow of liquid water from the soil to the leaf is proportional to water potential differences, the flow of water vapor from inside the leaf to the leaf surface is proportional to water vapor pressure differences. At the interface, a near-equilibrium exists between the liquid water and the water vapor in contact with it. Although different processes are used to describe water movement in different parts of the system, the flow of water is continuous across these discontinuities of processes.

A very small amount of the water that passes from the soil to the atmosphere through the plant remains in the plant as storage as the plant grows (less than 5%). An even smaller amount of the hydrogen and oxygen in the water taken up from the soil remains in the plant in the form of organic molecules synthesized by the plant.

Water that exits the stomata must pass through a thin, still-air layer adjacent to the leaf, called the leaf boundary layer, and be transported through the canopy space by turbulent mixing along the path of decreasing water vapor pressure. Ultimately the water vapor exits the plant canopy, passes through the planetary boundary layer in the lower few thousand meters of the atmosphere, and is lifted high into the atmosphere, where it condenses into clouds and eventually falls back to the Earth as precipitation to repeat the cycle.

Richards Equation

Transport of soil water affects heat and solute transport in soils, defines rates of biological processes in soil and water supply to plants, governs transpiration and ground water replenishment, controls runoff, and has many other important functions in the environment. Therefore, simulations of water transport in soil have many applications in hydrology, meteorology, agronomy, environmental protection, and other soil-related disciplines. Success of a multitude of projects depends on the correctness of the model of soil water transport.

The Richards' equation is the most often used model. It has been introduced by Richards 1931 who has suggested that the Darcy's law originally devised for saturated flow in porous media is also applicable to unsaturated flow in porous media. One-dimensional horizontal soil columns present the simplest systems to assess the validity of the Richards' equation. For such systems, Richards' equation reduces to,

$$\frac{\partial \theta}{\partial t} = \frac{\partial}{\partial x}\left[D(\theta)\frac{\partial \theta}{\partial x}\right]$$

Here u is the volumetric soil water content (m3 m23), D is the soil water diffusivity (m2 s 21), x is the distance from one of the ends of the column (m), t is time (s). Soil bulk density changes and soil water hysteresis are ignored in this formulation. Introduction of the Boltzmann variable,

$$\lambda = \frac{x}{t^{0.5}}$$

transforms equation ($\frac{\partial \theta}{\partial t} = \frac{\partial}{\partial x}\left[D(\theta)\frac{\partial \theta}{\partial x}\right]$) into an ordinary differential equation,

$$-\frac{\lambda}{2}\frac{d\theta}{d\lambda} = \frac{d}{d\lambda}\left[D(\theta)\frac{d\theta}{d\lambda}\right]$$

which has been used to find analytical solutions for soil water flow problems and also to find the dependence of the diffusivity D on soil water content θ. If equation ($-\frac{\lambda}{2}\frac{d\theta}{d\lambda} = \frac{d}{d\lambda}\left[D(\theta)\frac{d\theta}{d\lambda}\right]$) is applicable then soil water content is a function of the Boltzmann variable l, and, for the same values of soil water content, one should expect the same values of the Boltzmann variable.

Validity of Eq. ($-\frac{\lambda}{2}\frac{d\theta}{d\lambda} = \frac{d}{d\lambda}\left[D(\theta)\frac{d\theta}{d\lambda}\right]$) can be tested with experimental data consisting of observed soil moisture changes during infiltration in horizontal soil columns with initially uniform soil water content as shown in figure below. Distances and times at which the same values of water content have been observed must obey equations,

$$\frac{x_1}{t_1^{0.5}} = \frac{x_2}{t_2^{0.5}} = \frac{x_3}{t_3^{0.5}} = \cdots,$$

and, in general,

$$x = At^{0.5}$$

where the multiplier A depends only on water content. This equation means that the dependence between lg(x) and lg(t) plotted in log–log coordinates is linear and the slope of this dependence is 0.5 whereas the intercept depends on the water content.

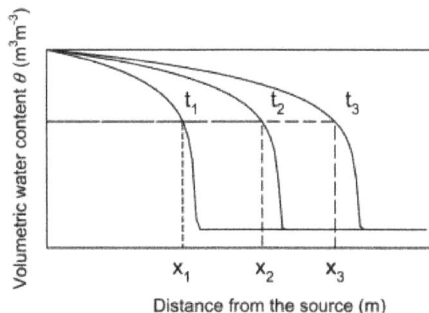

Figure: Sequential observation of soil water contents in an experiment on infiltration in horizontal soil columns

Significant deviations from equation ($x = At^{0.5}$) have been observed in many published experiments. Gardner and Widtsoe 1921 and Nielsen recorded the progress of the wetting front in air-dry soil uniformly packed in horizontal columns. A negative pressure head was held at one end of the columns. The largest distance where the wetting front was observed was 50 cm. Relationships between wetting front positions and time are shown in figure below. A linear dependence

$$\lg x = \lg A + q \lg t$$

can be traced in those figures. The companion Table below contains values of the slope q; all are significantly less than the value of 0.5 predicted by Eq. ($x = At^{0.5}$). Rawlins and Gardner 1963 and Ferguson and Gardner 1963 studied movement of water in horizontal soil columns using gamma ray attenuation. The furthest distance where water content was measured was 40 cm. Figure shows their data on the distance at which a particular water content had been reached and time when this water content was reached. Slopes of the regression lines at this figure are given in table below. Those slopes are also significantly less than 0.5. Similar data were published by Biggar and Taylor 1959 and Guerrini and Swartzendruber. On the other hand, Selim and Whisler reported data from experiments on horizontal infiltration in soil columns in which no significant deviations from equation ($x = At^{0.5}$) were found.

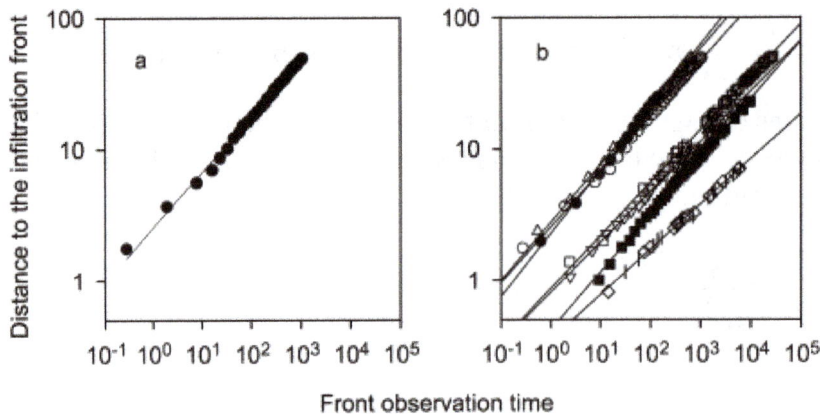

Front observation time

The objective of this work was to develop a physics-based model of water transport in unsaturated oils that would explain and simulate both deviations from and agreeing to the Boltzmann scaling.

Table: Values of the parameter q found from data on horizontal movement of water to soil columns

Soil	q (mean ± standard error)
Name and texture not reported	0.417 ± 0.006
Columbia silt loam wet at − 50 mb	0.402 ± 0.003
Columbia silt loam wet at − 100 mb	0.425 ± 0.006
Columbia silt loam wet with oil at − 2 mb	0.480 ± 0.008
Columbia silt loam wet with oil at − 38 mb	0.440 ± 0.003
Hesperia sandy loam at − 2 mb	0.440 ± 0.004
Hesperia sandy loam at − 50 mb	0.384 ± 0.002
Hesperia sandy loam wet at − 100 mb	0.344 ± 0.003

Soil	q (mean ± standard error)
Salkum silty clay loam, $\theta = 0.51$	0.439 ± 0.007
Salkum silty clay loam, $\theta = 0.50$	0.430 ± 0.008
Salkum silty clay loam, $\theta = 0.48$	0.437 ± 0.011
Salkum silty clay loam, $\theta = 0.45$	0.467 ± 0.009
Salkum silty clay loam, $\theta = 0.40$	0.479 ± 0.003
Salkum silty clay loam, $\theta = 0.05$	0.461 ± 0.002
Salkum silty clay loam, $\theta = 0.05$	0.454 ± 0.002
Salkum silty clay loam, $\theta = 0.10$	0.453 ± 0.002
Salkum silty clay loam, $\theta = 0.15$	0.452 ± 0.003
Salkum silty clay loam, $\theta = 0.20$	0.452 ± 0.003
Salkum silty clay loam, $\theta = 0.25$	0.452 ± 0.003
Salkum silty clay loam, $\theta = 0.30$	0.454 ± 0.003
Salkum silty clay loam, $\theta = 0.35$	0.458 ± 0.004
Salkum silty clay loam, $\theta = 0.40$	0.465 ± 0.006

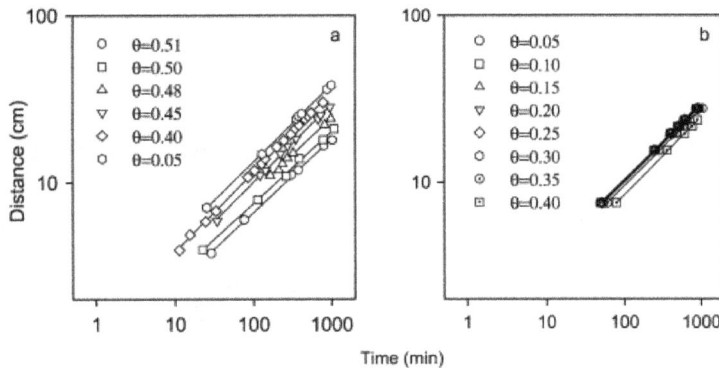

The Generalized Richards' Equation

Physical Model

For a long time, the Richards' equation first was considered to be a purely empirical flow equation, the result of combining the equation of continuity with the experimentally based 'Buckingham–Darcy' flux law. Bhattacharya et al. 1976 showed that this equation can be derived from physically based molecular assumption. These authors had to assume that water moves in a Brownian motion in the form of quasi-molecules. Assumption that particles perform a Brownian motion has been also instrumental in deriving the diffusion equation and a convective–dispersive equation. The Boltzmann scaling is applicable to both horizontal water transport equation and to the solute diffusion equation.

Solute transport in porous media has been shown to exhibit deviations from Boltzmann scaling and follow a more general scaling law with both q > 0.5 and q < 0.5. The physical model for such transport is the movement of particles that do not perform a Brownian motion because of constraints imposed either by structure of porous media or by the solute surface interactions. The case of q > 0.5 was interpreted resulting from Le´vy motion of solute particles, which is similar to the Brownian motion except that the relatively large transitions occur relatively more often. This may happen because of the presence of highly conductive fractures, channels, or macropores. The case q < 0.5 is also interpreted as non-Brownian transport of particles that remain motionless for extended periods of time, for example, when waiting periods have a power law distribution. Such a physical model was envisaged by Donald Nielsen and colleagues in 1962 who had suggested that the exponent q < 0.5 might occur because the infiltration front underwent 'jerky movements', i.e. immobility of the wetting front occurred for substantially extended time periods. This physical model of water particles being randomly trapped and having a power law distribution of waiting periods is used in this work to generalize the Richards' equation.

Mathematical Model

Particles with a power law distribution of waiting periods have an infinite mean waiting time, and it has been suggested to simulate the transport of an ensemble of such particles using fractional derivative on time. The transport equation appears to be similar to equation $\frac{\partial \theta}{\partial t} = \frac{\partial}{\partial x}\left[D(\theta)\frac{\partial \theta}{\partial x}\right]$ except that fractional derivative of water content on time is used:

$$\frac{\partial^{\gamma}\theta}{\partial t^{\gamma}} = \frac{\partial}{\partial x}\left[D\gamma(\theta)\frac{\partial \theta}{\partial x}\right]$$

Here g is the order of the fractional derivative, and Dg is fractional diffusivity. The order of the fractional derivative is less or equal to 1. Eq. ($Q = -D_{\gamma}(\theta)\frac{\partial \theta}{\partial x}$) transforms into the classical Richards' equation when g ¼ 1: The fractional diffusivity Dg depends on water content. The water flux Q is governed by the Darcy law

$$Q = -D_{\gamma}(\theta)\frac{\partial \theta}{\partial x}$$

as in the classical Richards' equation.

The meaning of a fractional derivative may be perceived from its finite-difference approximation. If the function u(t) is defined at time moments $t_0 = 0, t_1 = \Delta t, t_2 = 2\Delta t, ... t_n = n\Delta t$; and it is smooth at $t = 0$; then the following approximation can be used.

$$\frac{\partial^{\gamma}\theta}{\partial t^{\gamma}}\bigg|_{t=t_n}$$

$$\approx \frac{\theta^n - c_1\theta^{n-1} - c_2\theta^{n-2} - c_3\theta^{n-3} - \cdots - c_{n-1}\theta^1 - c_n\theta^0}{(\Delta t)^{\gamma}}$$

Here θ_n is the value of θ at time tn, un21 is the value of θ at time t_n, θ^{n-1}, etc. coefficients $c_j, j = 1,2,3,....$ depend on the order of the fractional differentiation as:

$$c_j = (-1)^{j-1}\binom{\gamma}{j} = \frac{\gamma(\gamma-1)\cdots(\gamma-j)}{1\cdot 2\cdots\cdot j}$$

The important feature of the fractional derivative approximation ($\approx \frac{\theta^n - c_1\theta^{n-1} - c_2\theta^{n-2} - c_3\theta^{n-3} - \cdots - c_{n-1}\theta^1 - c_n\theta^0}{(\Delta t)^{\gamma}}$) is the incorporation of the values of the dependent variable u not only at the current and the previous time moments t_n and t_{n-1}, as it is usually done with the first-order derivatives, but at all previous incremental time moments $t_n, t_{n-1}, t_{n-2},\dots t_1, t_0$.

Coefficients are decaying functions of j. Beginning from j ¼ 10; dependencies of c_j on j can be approximated by power laws $c = c_{10} j^{-m}$ where $m = 1 + \gamma$

By introducing an analog j of the Boltzmann variable as,

$$\xi = \frac{x}{t^{\lambda/2}},$$

Eq. ($Q = -D_\gamma(\theta)\dfrac{\partial \theta}{\partial x}$) can be easily transformed into,

$$\frac{\Gamma(1-\gamma/2)}{\Gamma(1-3\gamma/2)} \xi \frac{d\theta}{d\xi} = \frac{d}{d\xi}\left[d_\gamma \frac{d\theta}{d\xi}\right]$$

which reduces to Eq. ($-\dfrac{\lambda}{2}\dfrac{d\theta}{d\lambda} = \dfrac{d}{d\lambda}\left[D(\theta)\dfrac{d\theta}{d\lambda}\right]$) when $\gamma = 1$ (the derivation is in Appendix A, Γ is the gamma-function). Eq. ($\dfrac{\Gamma(1-\gamma/2)}{\Gamma(1-3\gamma/2)} \xi \dfrac{d\theta}{d\xi} = \dfrac{d}{d\xi}\left[d_\gamma \dfrac{d\theta}{d\xi}\right]$) shows that soil water content is a function of the variable ξ , and, for the same values of soil water content, one should expect the same values of the variable ξ . Therefore, the equation,

$$x = A(\theta) t^{\gamma/2}$$

or its analog

$$\lg x = \lg A + \frac{\gamma}{2} \lg t$$

has to be valid in experiments with horizontal infiltration in a soil column. Comparison of Eqs. ($\lg x = \lg A + \frac{\gamma}{2}\lg t$) and ($\lg x = \lg A + q\lg t$) leads to the conclusion that Eq. ($\lg x = \lg A + \frac{\gamma}{2}\lg t$) is indeed valid in experiments with horizontal infiltration and the empirical parameter q in this equation is equal to $\gamma/2$. Values of q in table above are all less than 0.5 which means that Eq. ($Q = -D_\gamma(\theta)\dfrac{\partial \theta}{\partial x}$) is valid in this experimental conditions where $\gamma < 1$.

Eq. ($\dfrac{\Gamma(1-\gamma/2)}{\Gamma(1-3\gamma/2)} \xi \dfrac{d\theta}{d\xi} = \dfrac{d}{d\xi}\left[d_\gamma \dfrac{d\theta}{d\xi}\right]$) can be rearranged to compute the fractional diffusivity $D\gamma$ from

experimentally defined. dependence of ξ on θ .

$$D(\theta) = \frac{\Gamma(1-\gamma/2)}{\Gamma(1-3\gamma/2)} \frac{d\xi}{d\theta} \int_{\theta_1}^{\theta} \xi(\theta) d\theta$$

where θ_i is the initial water content, when $\gamma = 1$; this equation reduces to Philip's 1955 equation to compute the hydraulic diffusivity:

$$D(\theta) = -\frac{1}{2}\frac{d\gamma}{d\theta}\int_{\theta_i}^{\theta}\lambda(\theta)d\theta$$

Water transport equations have to be solved numerically when boundary conditions are variable, the initial water content is not homogeneous, and the flow domain cannot be assumed semi-infinite. Using Eq. ($\approx \dfrac{\theta^n - c_1\theta^{n-1} - c_2\theta^{n-2} - c_3\theta^{n-3} - \cdots - c_{n-1}\theta^1 - c_n\theta^0}{(\Delta t)^{\gamma}}$), one derives an implicit finite difference

approximation of Eq. ($Q = -D_\gamma(\theta)\dfrac{\partial\theta}{\partial x}$) as,

$$\frac{\theta_i^n - \sum_{j=1}^{n} c_j\theta_i^{n-j}}{(\Delta t)^{\gamma}} = \frac{\frac{1}{2}\left[D(\theta_{i+1}^n) + D(\theta_i^n)\right]\frac{\theta_{i+1}^n - \theta_i^n}{\Delta x} - \frac{1}{2}\left[D(\theta_i^n) + D(\theta_{i-1}^n)\right]\frac{\theta_i^n - \theta_{i-1}^n}{\Delta x}}{\Delta x}$$

where the subscript i denotes values of θ at x= $x_i = i\Delta x$ $i = 1,2,....M-1$ Simple transformation converts this system of equations into a tridiagonal system on non-linear equations with respect to θ_i^n $i = 1,1,2,....M-1$,

$$\frac{D(\theta_{i-1}^n) + D(\theta_i^n)}{2(\Delta x)^2}\theta_{i-1}^n$$

$$-\left[\frac{D(\theta_{i-1}^n) + 2D(\theta_i^n) + D(\theta_{i+1}^n)}{2(\Delta x)^2} + \frac{1}{(\Delta t)^{\gamma}}\right]\theta_{i+1}^n$$

$$+\frac{D(\theta_{i-1}^n) + D(\theta_i^n)}{2(\Delta x)^2}\theta_{i+1}^n = -\frac{\sum_{j=1}^{n} c_j\theta_i^{n-j}}{(\Delta t)^{\gamma}}$$

for $i = 1,2,3,....M$: Two more equations are derived from boundary conditions, after that the resulting system of $M+1$ equations can be solved using iterations with Gauss elimination. The code to implement this algorithm for numerical solution of Eq. ($Q = -D_\gamma(\theta)\dfrac{\partial\theta}{\partial x}$) has been written in FORTRAN and is available from the corresponding author upon request. It was tested to see how accurately Eq. ($\lg x = \lg A + \dfrac{\gamma}{2}\lg t$) would apply to the numerical solution. Function $D_\gamma(\theta)$ was taken as exp $(5(\theta - 1.5)$ boundary conditions were $\theta_0^n = 1$ and $\theta_{M-1}^n = \theta_M^n$ for $n = 0,1,2,....$, the initial condition was $\theta_i^0 = 0$ $i = 1,2,...,M$; intervals for x and t were $0 \leq x \leq 50 cm, 0 \leq t \leq 2000 \min$, the discretization was made with: The scaling ($\lg x = \lg A + \dfrac{\gamma}{2}\lg t$) was applicable to the numerical solution with 2% difference between γ values from the numerical solution and γ values assumed to obtain this solution for $0 \leq x \leq 30$ cm.

Apparent Scale Effects on Hydraulic Diffusivity

The dependencies of Boltzmann variable λ and the fractional scaling variable ξ on volumetric water content θ were found from data of Ferguson and Gardner 1963 collected at 3.5, 7.5, 15.5 and 27.8 cm from the source. The experimental dependencies of the Boltzmann variable on θ for those distances are shown in figure below Dependencies are distinctly different for different depths

To compute the hydraulic diffusivity from the Philips' equation ($D(\theta) = -\dfrac{1}{2}\dfrac{d\gamma}{d\theta}\int_{\theta_i}^{\theta}\lambda(\theta)d\theta$), we had to approximate the dependencies shown in figure below because the equation includes derivatives of the dependence of on λ and θ. Data for each depth were fitted with the empirical logistic equation

$$\lambda = \lambda_0 \frac{1 + u\exp(v\theta_0)}{1 + u\exp(v\theta)}\exp[v(\theta - \theta_0)]$$

$$+\lambda_{max}\frac{1 - \exp[v(\theta - \theta_0)]}{1 + u\exp(v\theta_0)}$$

Where θ_0 is the maximum water content close to porosity where observation are available, λ_0 is the observed value of λ at $\theta = \theta_0$, λ_{max}, u and v are fitting parameters. Lines in figure (a) below show results of this fitting. Dependencies of hydraulic diffusivity on water contents computed according to Eq. ($D(\theta) = -\dfrac{1}{2}\dfrac{d\gamma}{d\theta}\int_{\theta_i}^{\theta}\lambda(\theta)d\theta$) are shown in Figure of volumetric water content. Differences among dependencies of λ on θ at different depths cause variability in diffusivity values. The average range of variations is about half an order of magnitude for any given θ value.

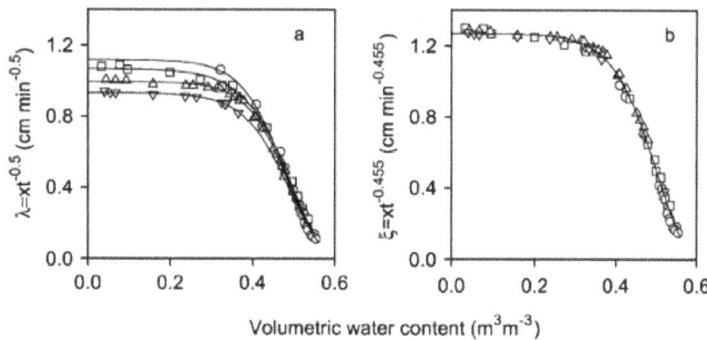

Figure: Boltzmann scaling variable λ and fractional scaling ξ as functions of the volumetric water contents from observatins of Ferguson and Gardner at several distances from the infiltration source; o—3.5 cm, □—7.5 cm, Δ—15.5 cm, ∇—27.5 cm

Dependencies of the fractional scaling variable ξ on θ are shown in figure (b) above. The order of the fractional derivative γ was taken as $\bar{q}/2$ where $\bar{q}$ =0:455 is the average value for this data set in table above. Data for different depths coalesce, and it is possible to derive a unique, depth-independent function $\xi(\theta)$ This function could be used to generate a unique dependence of the fractional diffusivity on u as shown in figure below.

We carried out numerical experiments consisting of water transport simulations with the generalized Richards' equation and calculating classic Richards diffusivity from water content profiles at several times. Results of one such experiment are shown in figure. The order of the fractional derivative was g ¼ 0.8, the initial water content and the porosity were 0.04 and 0.55 m3 m23 , respectively, the diffusivity was $D_\gamma = 40\theta^{(1.8 - 2.5 In\theta)}cm\,min^{-0.8}$; the distances and time ranged from 0 to 40 cm and from 0 to 1200 min, respectively. Figure (a) below shows dependencies of the Boltzmann variable on water content obtained from water content profiles at different times. The decrease in values of λ as time progresses can be observed. This decrease causes an apparent decrease in diffusivity values obtained for the same water contents as time increases. When the water contents at different distances rather than times were used to compute λ, the diffusivity decreased with the distance (data not shown).

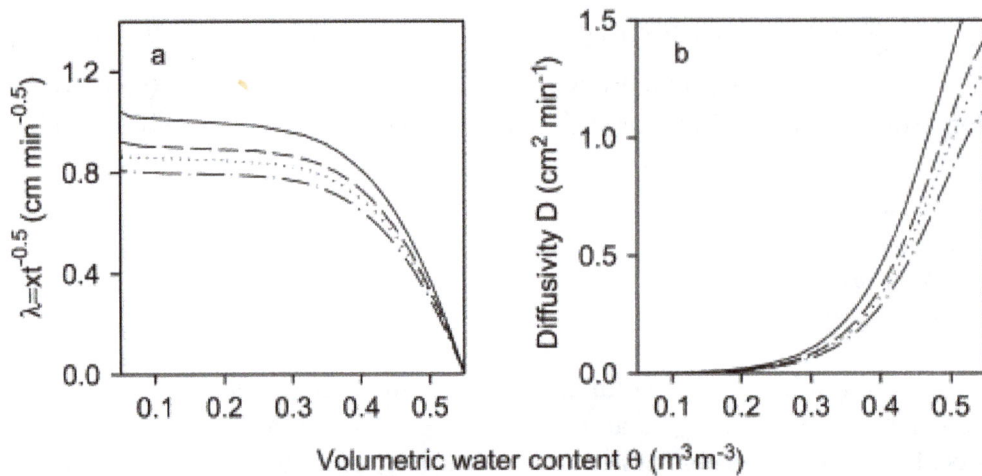

Figure: Boltzmann scaling variable l and diffusivity from the numerical solution of the generalized Richards' equation for several simulated times; 120 min, 360 min, 600 min, 1200 min.

Water Content

The moisture content of soil also referred to as water content is an indicator of the amount of water present in soil. By definition, moisture content is the ratio of the mass of water in a sample to the mass of solids in the sample, expressed as a percentage. In equation form,

$$w = \frac{M_w}{M_g} \times 100$$

where:

w = moisture content of soil expressed as a percentage

M_w = mass of water in soil sample i.e., initial mass of moist soil minus mass of oven-dried soil

M_g = mass of soil solids in sample i.e., the soil's "oven-dried mass"

M_w and M_g may be expressed in any units of mass, but both should be expressed in the same unit.

It might be noted that the moisture content could be mistakenly de fined as the ratio of mass of water to total mass of moist soil rather than to the mass of oven-dried soil. Because the total mass of moist soil is the sum of the mass of water and oven-dried soil, this incorrect definition would give a fraction in which both numerator and denominator vary but not in the same proportion according to the amount of moisture present. Such a definition would be undesirable, because moisture con tent would then be based on a varying quantity of moist mass of soil rather than a constant quantity of oven-dried soil. Stated another way, with the incorrect definition, the moisture content would not be directly proportional to the mass of water present. With the correct definition given by Eq. 3-1, moisture content is directly proportional to the mass of water present. This characteristic makes moisture con tent, as defined by Eq. 3-1, one of the most useful and important soil parameters.

Volumetric water content, θ, is defined mathematically as:

$$\theta = \frac{V_w}{V_{wet}}$$

where V_w is the volume of water and $V_{wet} = V_5 + V_w + V_a$ is equal to the total volume of the wet material, i.e. of the sum of the volume of solid host material (e.g., soil particles, vegetation tissue) V_s, of water V_w, and of air V_a.

Gravimetric water content is expressed by mass (weight) as follows:

$$u = \frac{m_w}{m}$$

where m_w is the mass of water and m is the mass of the substance. Normally the latter is taken before drying:

$$u' = \frac{m_w}{m_{wet}}$$

except for woodworking, geotechnical and soil science applications where oven-dried material is used instead:

$$u'' = \frac{m_w}{m_{dry}}$$

To convert gravimetric water content to volumetric water content, multiply the gravimetric water content by the bulk specific gravity SG of the material:

$$\theta = u \times SG.$$

Derived Quantities

In soil mechanics and petroleum engineering the water saturation or degree of saturation, S_w is defined as

$$S_w = \frac{V_w}{V_v} = \frac{V_w}{V\phi} = \frac{\theta}{\phi}$$

where $\phi = V_v / V$ is the porosity, in terms of the volume of void or pore space V_v and the total volume of the substance V. Values of S_w can range from 0 (dry) to 1 (saturated). In reality, S_w never reaches 0 or 1 - these are idealizations for engineering use.

The normalized water content, Θ, (also called effective saturation or S_e) is a dimensionless value defined by van Genuchten as:

$$\Theta = \frac{\theta - \theta_r}{\theta_s - \theta_r}$$

where θ is the volumetric water content; θ_r is the residual water content, defined as the water content for which the gradient $d\theta / dh$ becomes zero; and, θ_s is the saturated water content, which is equivalent to porosity, ϕ.

Measurement

Direct Methods

Water content can be directly measured using a known volume of the material, and a drying oven. Volumetric water content, θ, is calculated via the volume of water θ_s and the mass of water:

$$V_w = \frac{m_w}{\rho_w} = \frac{m_{wet} - m_{dry}}{\rho_w}$$

where

m_{wet} and m_{dry} are the masses of the sample before and after drying in the oven;

ρ_w is the density of water; and

For materials that change in volume with water content, such as coal, the water content, u, is expressed in terms of the mass of water per unit mass of the moist specimen:

$$u' = \frac{m_{wet} - m_{dry}}{m_{wet}}$$

However, geotechnics requires the moisture content to be expressed with respect to the sample's dry weight (often as a percentage, i.e. % moisture content = $u \times 100\%$)

$$u'' = \frac{m_{wet} - m_{dry}}{m_{dry}}$$

For wood, the convention is to report moisture content on oven-dry basis (i.e. generally drying sample in an oven set at 105 deg Celsius for 24 hours). In wood drying, this is an important concept.

Laboratory Methods

Other methods that determine water content of a sample include chemical titrations (for example the Karl Fischer titration), determining mass loss on heating (perhaps in the presence of an inert gas), or after freeze drying. In the food industry the Dean-Stark method is also commonly used.

From the Annual Book of ASTM (American Society for Testing and Materials) Standards, the total evaporable moisture content in Aggregate (C 566) can be calculated with the formula:

$$p = \frac{W - D}{W}$$

where p is the fraction of total evaporable moisture content of sample, W is the mass of the original sample, and D is mass of dried sample.

Field capacity is a measurement that has to do with the ability of soil in a given area to absorb water, once all excess and surface water has been drained from the area. Assessing the field capacity of soil is very important in determining the type of crops to grow in a particular section of land, as well as judging the capacity of that land to support buildings of various types. The result of these assessments is usually presented as a percentage.

Calculating field capacity is a process that normally takes a couple of days. The soil is saturated to the point that there is some water left standing on the ground surface. The standing water is removed, then the remaining water is allowed to seep into the soil and eventually drain away. After anywhere from twenty-four to forty-eight hours, the moisture content of the soil is tested. This process makes it possible to get a good idea of how much moisture the soil can retain while still remaining viable for planting or as a site for constructions.

One of the benefits of testing a tract of land using this process is that it can help growers to determine what types of crops to plant in the area. Depending on the actual soil moisture field capacity, it may be advantageous to go with crops that require less water retention in the soil in order to thrive. High water content will indicate that the soil is better suited for planting crops that require a great deal of moisture in order to grow properly. A calculation of this type is often known as a field capacity wilting point, since the idea is to determine what plants will grow and not wilt and decay due to exposure to the higher moisture content in the soil.

Determining the field capacity is also helpful when planning construction in an area. Soil that can retain a great deal of moisture and still remain solid is often a good choice for construction ranging from storage facilities to multi-story buildings. In this application, the soil analysis helps builders to determine the best approach to laying a foundation and making sure the building with not shift and crack due to the average amount of ground saturation that occurs through the calendar year. Taking the time to run a field capacity test before erecting any kind of building will enhance the chances of the building being safe and useful for many years without the need to shore up a failing foundation.

Field Capacity

Field capacity is a measurement that has to do with the ability of soil in a given area to absorb water, once all excess and surface water has been drained from the area. Assessing the field capacity of soil is very important in determining the type of crops to grow in a particular section of land, as well as judging the capacity of that land to support buildings of various types. The result of these assessments is usually presented as a percentage.

Calculating field capacity is a process that normally takes a couple of days. The soil is saturated to the point that there is some water left standing on the ground surface. The standing water is removed, and then the remaining water is allowed to seep into the soil and eventually drain away.

After anywhere from twenty-four to forty-eight hours, the moisture content of the soil is tested. This process makes it possible to get a good idea of how much moisture the soil can retain while still remaining viable for planting or as a site for constructions.

One of the benefits of testing a tract of land using this process is that it can help growers to determine what types of crops to plant in the area. Depending on the actual soil moisture field capacity, it may be advantageous to go with crops that require less water retention in the soil in order to thrive. High water content will indicate that the soil is better suited for planting crops that require a great deal of moisture in order to grow properly. A calculation of this type is often known as a field capacity wilting point, since the idea is to determine what plants will grow and not wilt and decay due to exposure to the higher moisture content in the soil.

Determining the field capacity is also helpful when planning construction in an area. Soil that can retain a great deal of moisture and still remain solid is often a good choice for construction ranging from storage facilities to multi-story buildings. In this application, the soil analysis helps builders to determine the best approach to laying a foundation and making sure the building with not shift and crack due to the average amount of ground saturation that occurs through the calendar year. Taking the time to run a field capacity test before erecting any kind of building will enhance the chances of the building being safe and useful for many years without the need to shore up a failing foundation.

References

- T. William Lambe & Robert V. Whitman (1969). "Chapter 3: Description of an Assemblage of Particles". Soil Mechanics (First ed.). John Wiley & Sons, Inc. p. 553. ISBN 0-471-51192-7.

- Soil-Plant-Atmoshere-Contin: ecosensing.org, Retrieved 27 May 2018

- van Genuchten, M.Th. (1980). "A closed-form equation for predicting the hydraulic conductivity of unsaturated soils". Soil Science Society of America Journal. 44 (5): 892–898. Bibcode:1980SSASJ..44..892V. doi:10.2136/sssaj1980.03615995004400050002x.

- What-is-field-capacity: wisegeek.com, Retrieved 22 June 2018

- Lawrence, J. E. & G. M. Hornberger (2007). "Soil moisture variability across climate zones". Geophys. Res. Lett. 34 (L20402): L20402. Bibcode:2007GeoRL..3420402L. doi:10.1029/2007GL031382.

Natural Stream Processes

Rivers and streams provide habitats for aquatic plants and animals. They also carry sediment and water from high elevated areas to downstream estuaries, lakes and oceans. The various aspects integral to natural stream processes include stream flow, stream stability, stream restoration, etc. These have been extensively discussed in this chapter.

Streams and rivers are integral parts of the landscape that carry water and sediment from high elevations to downstream lakes, estuaries, and oceans. The land area draining to a stream or river is defined as its watershed. When rain falls in a watershed, it either runs off the land surface, infiltrates into the soil, or evaporates. As surface runoff moves downslope, it concentrates in low areas and forms small stream channels. These are referred to as ephemeral channels that only carry water during rainfall runoff. Downstream from ephemeral channels are intermittent streams, which carry water during wet times of the year. These streams are partially supplied by groundwater rising to the surface as stream baseflow. They dry up when groundwater levels drop. Further downstream where baseflow is large enough to sustain stream flow throughout the year, perennial streams are formed. The size and flow of a stream are directly related to its watershed area. Other factors which affect channel size and stream flow are land use, soil types, topography, and climate. The morphology, or size and shape, of the channel reflect all of these factors.

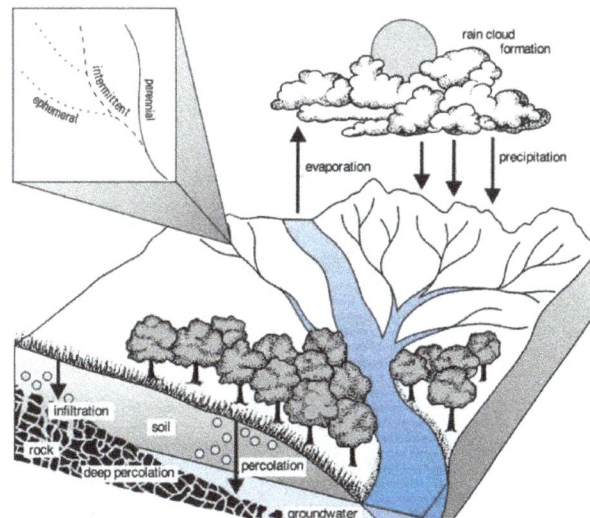

Figure: Hydrologic Cycle showing rainfall, runoff, infiltration, groundwater flow, and stream network.

While streams and rivers vary greatly in size, shape, slope, and bed materials, all streams share common characteristics. Streams have left and right streambanks (looking downstream) and streambeds consisting of mixtures of bedrock, boulders, cobble, gravel, sand, or silt/clay. Other physical characteristics shared by some stream types include pools, riffles, steps, point bars, meanders, floodplains, and terraces. All of these characteristics are related to the interactions among climate, geology, topography, vegetation and land use of the watershed. (Each of these character-istics will be defined in this fact sheet.) The study of these interactions and the resulting streams and rivers is called fluvial geomorphology.

In addition to transporting water and sediment, natural streams also provide the habitat for many aquatic organisms including fish, amphibians, insects, mollusks, and plants. Trees and shrubs along the banks provide a food source and regulate water temperatures. Channel features like pools, riffles, steps, and undercut banks provide diversity of habitat, oxygenation, and cover. For these reasons natural resource managers increasingly use natural channel designs to restore im-paired streams.

Bankfull Stage and Discharge

The most important stream process in defining channel form is the bankfull discharge, which is sometimes referred to as the effective discharge, or dominant discharge. Bankfull discharge is the flow that transports the majority of a stream's sediment load over time and thereby forms the channel. The bankfull stage, during bankfull flow is the point at which flooding occurs on the floodplain. This may or may not be the top of the stream- bank. If the stream has downcut due to changes in the watershed or streamside vegetation, the floodplain stage may be a small bench or scour line on the streambank. In this case, the top of the bank, which was formerly the floodplain, is called a terrace. A stream with terraces close to the top of the banks is an in-cised, or entrenched stream. If the stream is not entrenched, then bankfull is near the top of the bank. On average, bankfull discharge occurs approximately every 1.5 years. In other words, each year there is about a 67 percent chance of having a bankfull streamflow event. The Rosgen stream classification system uses bankfull stage as the basis for measuring the width/depth ratio and entrenchment ratio, two of the most important delineative criteria. Therefore, it is critical to correctly identify bankfull stage when classifying streams and designing stream restoration measures.

Figure: Photograph of an incised stream showing bankfull stage,
developing floodplain, and terrace.

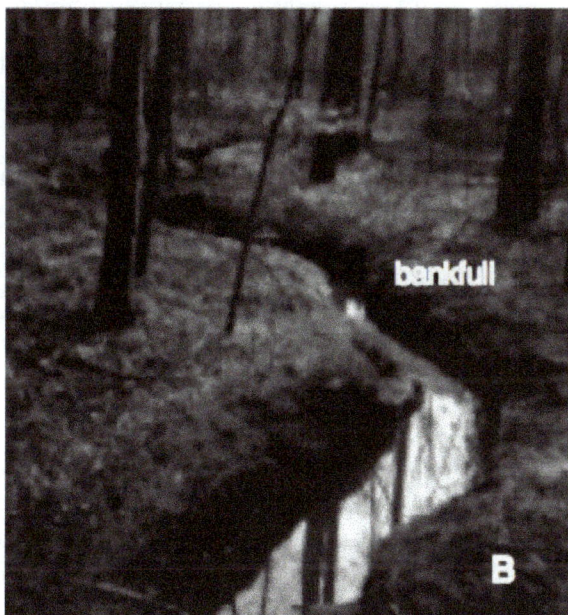

Figure: Photograph of a stream showing bankfull as the top of the bank

Natural Channel Stability

A naturally stable stream channel maintains its dimension, pattern, and profile over time so that the stream does not degrade or aggrade. Stable streams migrate across the landscape slowly over long periods of time while maintaining their form and function. Naturally stable streams must be able to transport the sediment load supplied by the watershed. Instability occurs when scouring causes the channel to incise (degrade) or excessive deposition causes the channel bed to rise (aggrade). A generalized relationship of stream stability is shown as a schematic drawing in figure below. The drawing shows that the product of sediment load and sediment size is proportional to the product of stream slope and discharge or stream power. A change in any one of these variables causes a rapid physical adjustment in the stream channel.

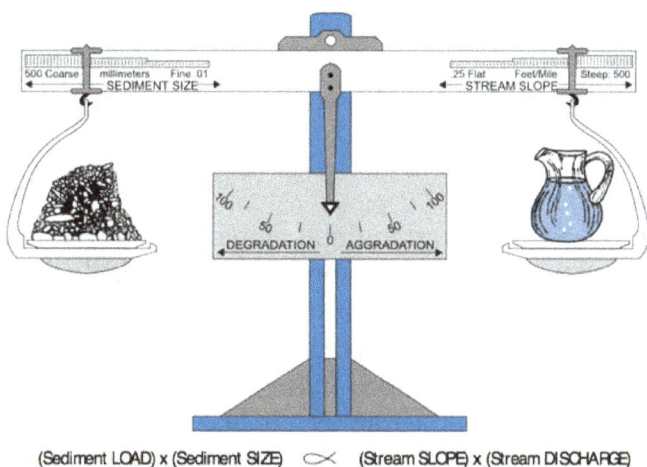

(Sediment LOAD) x (Sediment SIZE) ∝ (Stream SLOPE) x (Stream DISCHARGE)

Figure: Schematic drawing showing stream stability

Channel Dimension

The dimension of a stream is its cross-sectional area (width multiplied by mean depth). The width of a stream generally increases in the downstream direction in proportion to the square root of discharge. Stream width is a function of discharge (occurrence and magnitude), sediment transport (size and type), and the stream bed and bank materials. North Carolina has a humid subtropical climate with an abundance of vegetation and rainfall throughout the year. Vegetation along the streambanks provides resistance to erosion so our streams are often narrower than streams in more arid regions. The mean depth of a stream varies greatly from reach to reach depending on channel slope and riffle/pool or step/pool spacing.

Stream Pattern

Stream pattern describes the "plan view" of a channel as seen from above. Streams are rarely straight. They tend to follow a sinuous path across a floodplain. The sinuosity of a stream is defined as the channel length following the deepest point in the channel divided by the valley length. A meander increases resistance and reduces channel gradient relative to a straight reach. The meander geometry and spacing of riffles and pools adjust so that the stream per- forms minimal work. Stream pattern is qualitatively described as straight, meandering, or braided. Braided channels are less sinuous than meandering streams and possess three or more channels. Quantitatively, stream pattern can be defined through the following measurements shown in figure below.

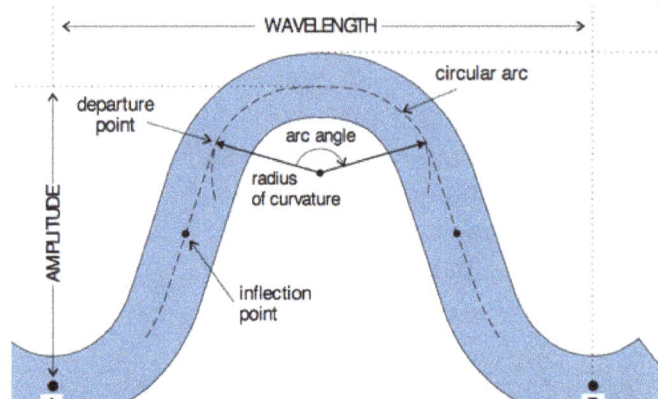

Figure: Meander wavelength, radius of curvature, amplitude, and belt width.

Stream Profile

The profile of a stream refers to its longitudinal slope. At the watershed scale, channel slope generally decreases in the downstream direction. The size of the bed material also decreases in the downstream direction. Channel slope is inversely related to sinuosity. This means that steep streams have low sinuosities and flat streams have high sinuosities. The profile of the streambed can be irregular because of variations in bed material size and shape, riffle/pool spacing, and other variables. The water surface profile mimics the bed profile at low flows. As water rises in a channel during storms, the water surface profile becomes more uniform as illustrated in figure below.

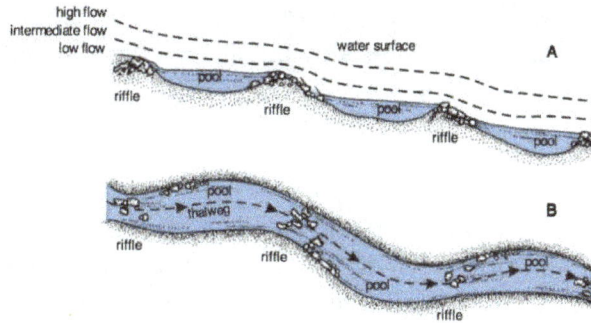

Figure: Bed and water surface slope at baseflow and stormflow

Channel Features

Natural streams have sequences of riffles and pools or steps and pools that maintain channel slope and stability. These features are shown in figure above. The riffle is a bed feature with gravel or larger size particles. The water depth is relatively shallow and the slope is steeper than the average slope of the channel. At low flows, water moves faster over riffles, which provides oxygen to the stream. Riffles are found entering and exiting meanders and control the streambed elevation. Pools are located on the outside bends of meanders between riffles. The pool has a flat slope and is much deeper than the average depth. At low flows, pools are depositional features and riffles are scour features. At high flows, however, the pool scours and bed material deposits on the riffle. This occurs because a force, called shear stress, applied to the streambed increases with depth and slope. Slope and depth increase rapidly over the pools during large storms, increasing shear stress and causing scour. The inside of the meander bend is a depositional feature called a point bar, which also helps maintain channel form.

Step/pool sequences are found in high gradient streams. Steps are vertical drops often formed by large boulders, bedrock knickpoints, downed trees, etc. Deep pools are found at the bottom of each step. The step provides grade control and the pool dissipates energy. The spacing of step pools gets closer as the channel slope increases.

Stream Flow

The volume of water in a stream moving downslope at a given time, known as streamflow or stream discharge, is a combination of surface-water runoff and baseflow. Streamflow varies as these component amounts vary.

Flow Within The River

As water flows over a surface, the surface creates some resistance, and water closest to the ground is slowed. As the depth of the water deepens, this slowing near the ground and increasing away from the streambed becomes more noticeable. Measurements made in a variety of streams shows that velocity increases as the logarithm of the distance from the bottom of the stream channel.

Where the soil has enough cohesiveness (either innately or because of root systems) to create walls along the stream's bank, these walls also create some resistance, so the velocity is slowed near the sides of the channel. The overall effect of this slowing at the edges is that the water in the center moves the fastest.

Measuring Streamflow

Flow in the river is measured as volume per unit of time. The traditional unit of measure in the United States is cubic feet per second (cfs), although measurements today are often in cubic meters per second. To calculate the flow (discharge) of a river at a given point in time, the cross-sectional area of a river is divided into segments of equal width. The depth and average velocity of each segment is then measured. Average velocity is approximately 0.6 of the depth; velocity also can be measured directly by a current (or flow) meter. The width multiplied by the depth and the average velocity of each segment produces the discharge (in volume per time) for that segment. The sum of the discharges of the individual segments is the discharge of the river.

Flood Events

Floods occur when a stream flows over its banks. Studies have identified the relationship between flood frequency and volume within a stream channel. Estimates are that a river reaches this "bank-full discharge" once every 1 to 2 years, on average.

Flood events, particularly flash floods, often occur as a consequence of a storm with heavy or sustained rainfall. Statistical and mathematical methods have been developed to estimate volume, lag time, and duration of a flood based on the size (magnitude) of a rainfall event. These estimates depend on the characteristics of an individual basin.

The volume of a given event can be estimated as the amount of rainfall minus the amount of water lost in saturating the soil, minus the amount lost through a combination of evaporation to the atmosphere and transpiration from plants. (The combination of evaporation and transpiration collectively are called evapotranspiration.)

The lag time between the peak of a storm event and the flood peak is based on drainage basin characteristics such as size, shape, gradient, presence of wetlands or lakes, and amount of impermeable surface (such as concrete and asphalt). The lag time is shorter in smaller, steeper drainages with impermeable surface, and longer in larger basins with forests, wetlands, and lakes. Also, duration of the flood is longer than the storm itself because it takes time for water falling in the basin to flow overland to reach streams and rivers. It also takes time for water in the channel itself to carry this water to the basin's outflow point, such as a confluence with a larger river.

When streams are at flood stage, some water moves from the stream into the streambank. Some of this water occurs as bank storage that flows directly back into the stream as its water level drops. If the stream is hydraulically connected to an aquifer, however, a significant amount of the water moves into the aquifer.

Stream Interaction with its Environment

As the river moves downgradient (downhill), it interacts with its environment, both geologic (i.e., rocks and soil) and organic (such as tree roots and large woody debris). When the stream is shallow and the debris large, the water moves around the obstacles. As the volume of water increases, and the river becomes deeper, it has a greater ability to erode, and begins to cut into the underlying soils and pick up smaller particles. When and where this happens depends on factors like the rock type (lithology), rock or sediment size, and the cohesiveness of the streambank materials (often provided by the root structure of streamside vegetation).

As a river flows over and around obstacles, sometimes picking up materials and other times dropping them, the flow is almost always turbulent. The possible exception is very near the boundaries of a slow-moving river with relatively smooth channel boundaries.

Human-induced changes to stream channels can greatly modify the streamflow characteristics. Channelizing or straightening a river by dredging can more efficiently pass flood discharges through the straightened segment, but the faster water velocity can erode more sediment. Restricting a stream within artificial embankments, such as floodwalls, prevents it from naturally meandering, and can create new problems. Any human modifications to stream channels and floodplains should take into account the physics, biology, and ecology of a stream.

Hyporheic Zone

Recently, increasing interest has been focused on hyporheic flow. This is water that flows beneath and adjacent to the river, interacting with both the water in the stream and the groundwater. The result is an exchange of waters of different characteristics (such as temperature), creating microclimates within the river, and providing extended refuge outside the channel for small organisms, all considered to be part of a larger stream system.

Stream Form

The interaction between water and Earth's surface results in some fairly commonly observed stream shapes. * In the mountains, when the water is shallow and the rocks large, the water moves around and over the obstacles. This is termed a cascade. Farther downslope, where sediment is available and water has enough energy to move the sediment through the river system, meanders (curves) may form. This is probably the shape most commonly associated with rivers. Many studies have taken place and many theories developed to try to explain the cause and characteristics of the meander, using mathematics, physics, and theories of conservation of energy. At the end of the system, where drainage is more developed, where sediment supply is high, and/or where the stream has less energy (perhaps as a result of a lower gradient), the river may become braided.

Stream Flow and Sediment Transport

Stream velocity is the speed of the water in the stream. Units are distance per time (e.g., meters per second or feet per second). Stream velocity is greatest in midstream near the surface and is slowest along the stream bed and banks due to friction.

Hydraulic radius (HR or just R) is the ratio of the cross-sectional area divided by the wetted perimeter. For a hypothetical stream with a rectangular cross sectional shape (a stream with a flat bottom and vertical sides) the cross-sectional area is simply the width multiplied by the depth (W * D). For the same hypothetical stream the wetted perimeter would be the depth plus the width plus the depth (W + 2D). The greater the cross-sectional area in comparison to the wetted perimeter, the more freely flowing will the stream be because less of the water in the stream is in proximity to the frictional bed. So as hydraulic radius increases so will velocity (all other factors being equal).

Stream discharge is the quantity (volume) of water passing by a given point in a certain amount of time. It is calculated as Q = V * A, where V is the stream velocity and A is the stream's cross-sectional area. Units of discharge are volume per time (e.g., m^3/sec or million gallons per day, mgpd).

At low velocity, especially if the stream bed is smooth, streams may exhibit laminar flow in which all of the water molecules flow in parallel paths. At higher velocities turbulence is introduced into the flow (turbulent flow). The water molecules don't follow parallel paths.

Streams carry dissolved ions as dissolved load, fine clay and silt particles as suspended load, and coarse sands and gravels as bed load. Fine particles will only remain suspended if flow is turbulent. In laminar flow, suspended particles will slowly settle to the bed.

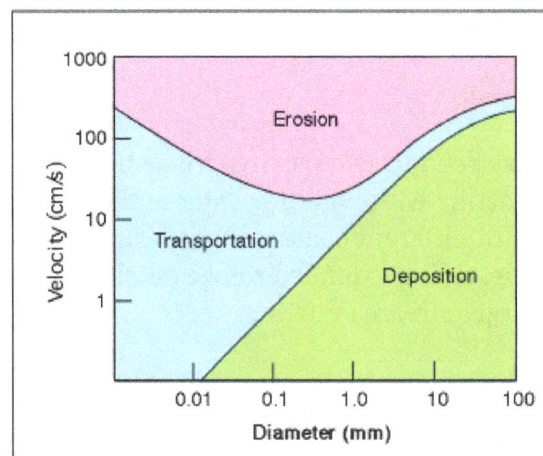

Hjulstrom's Diagram plots two curves representing 1) the minimum stream velocity required to erode sediments of varying sizes from the stream bed, and 2) the minimum velocity required to continue to transport sediments of varying sizes. Notice that for coarser sediments (sand and gravel) it takes just a little higher velocity to initially erode particles than it takes to continue to transport them. For small particles (clay and silt) considerably higer velocities are required for erosion than for transportation because these finer particles have cohesion resulting from electrostatic attractions. Think of how sticky wet mud is.

Stream competence refers to the heaviest particles a stream can carry. Stream competence depends on stream velocity (as shown on the Hjulstrom diagram above). The faster the current, the heavier the particle that can be transported.

Stream is the maximum amount of solid load (bed and suspended) a stream can carry. It depends on both the discharge and the velocity (since velocity affects the competence and therefore the range of particle sizes that may be transported).

As stream velocity and discharge increase so do competence and capacity. But it is not a linear relationship (e.g., doubling velocity and discharge do not simply double competence and capacity). Competence varies as approximately the sixth power of velocity. For example, doubling the velocity results in a 64 times increase in the competence.

Capacity varies as the discharge squared or cubed. So tripling the discharge results in a 9 to 27 times increase in the capacity.

Therefore, most of the work of streams is accomplished during floods when stream velocity and discharge (and therefore competence and capacity) are many times their level during low flow regimes. This work is in the form of bed scouring (erosion), sediment transport (bed and suspended loads), and sediment deposition.

Stream Dynamics

Perennial and Ephemeral Streams

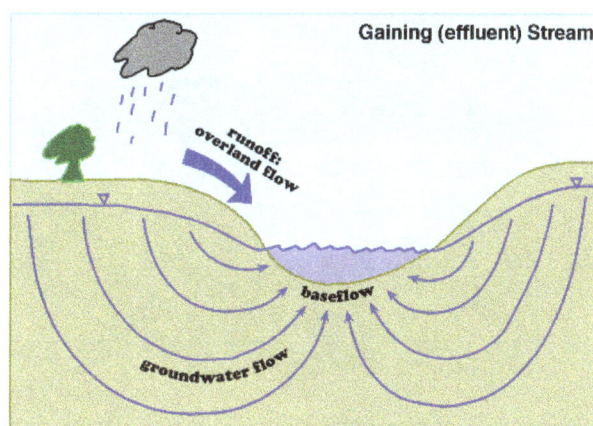

Gaining (effluent) streams: It receive water from the groundwater. In other words, a gaining stream discharges water from the water table. On the other hand losing (influent) streams lie above the water table (e.g., in an arid climate) and water seeps through the stream bed to recharge the water table below. Gaining streams are perennial streams: they flow year around. Losing streams are typically ephemeral streams: they do not flow year round. Th. only flow when there is sufficient runoff from recent rains or spring snowmelt. Some streams are gaining part of the year and losing part of the year or just in particular years, as the water table drops during an extended dry season.

Streams have two sources of water: storm charge, from overland flow after rain events, and base-flow, supplied by groundwater.

Flood Erosion and Deposition: As flood waters rise, the slope of the stream as it flows to its base level (e.g., the ocean or a lake) increases. Also, as stream depth increases, the hydraulic radius increases thereby making the stream more free flowing. Both of these factors lead to an increase in stream velocity. The increased velocity and the increased cross-sectional area mean that discharge increases. As discharge and velocity increase so do the stream's competence and capacity. In the rising stages of a flood much sediment is dumped into streams by overland flow and gully wash. This can result in some aggradation or building up of sediments on the stream bed. However, after the flood peaks less sediment is carried and a great deal of bed scouring (erosion) occurs. As the flood subsides and competence and capacity decline sediments are deposited and the stream bed aggrades again. Even though the stream bed may return to somewhat like its pre-flood state, huge quantities of sediments have been transported downstream. Much fine sediment has probably been deposited on the flood plain.

Stream Patterns

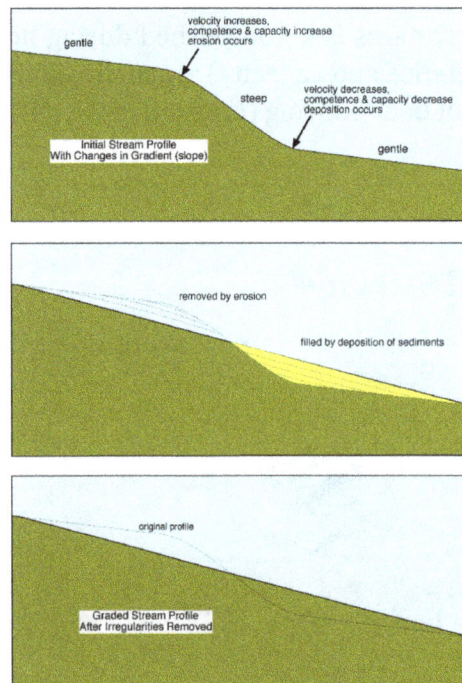

Meandering Streams: At a bend in a stream the water's momentum carries the mass of the water against the outer bank. Water piles up on the outer bank making it a little deeper and the inner bank a little shallower. The greater depth on the outer side of the bend also leads to higher velocity at the outer bank. The greater velocity combined with the greater inertial force on the outer bank erodes a deeper channel. The deeper channel reinforces the velocity increase. The inner bank remains shallower, increasing friction, thereby reducing the velocity.

Where the depth and velocity of the water on the outer bank increase so do the competence and capacity. Erosion occurs on the outer bank or cut bank. Where velocity of the water on the inner bank decreases so do the competence and capacity. Deposition occurs, leading to the formation of a point bar. Over time, the position of the stream changes as the bend migrates in the direction of the cut bank. As oxbow bends accentuate and migrate, two bends can erode together forming a cutoff and leaving an oxbow lake.

Graded Streams: Considering the longitudinal (downstream) profile of a stream:

Where a stream flows down a steep slope velocity will increase which will result in increased erosion. Where that stream then flows onto a gentler slope velocity decreases and deposition will result. This process will reduce the slope of steep stretches and increase the slope of flatter stretches resulting in a more even slope through the course of the stream.

The ideal graded profile of a stream is concave upward: steeper near the head or beginning and flatter near the bottom or mouth of the stream. The reason for this is that in the upper reaches of a stream its discharge is smaller. As streams merge with other streams their discharge increases, their cross-sectional area increases, and their hydraulic radius increases. As one goes downstream and the stream grows in size the waters flow more freely. In the upper reaches, a small stream must be steeper to transport its sediments.

The extra gravitational energy on the steeper slope is needed to overcome the frictional forces in the shallow stream. If the slope is too gentle and velocity is too slow to transport the sediments being supplied by weathering and erosion, the sediments will pile up. This increases the gradient which causes the water to flow faster which increases erosion and transport, which then reduces the gradient. In the lower reaches of a stream, where the discharge is greater, since friction is less the stream need not be so steep to transport the load. If it were steeper than needed to transport the sediments erosion would result. But this would decrease the gradient leading to a decrease in erosion.

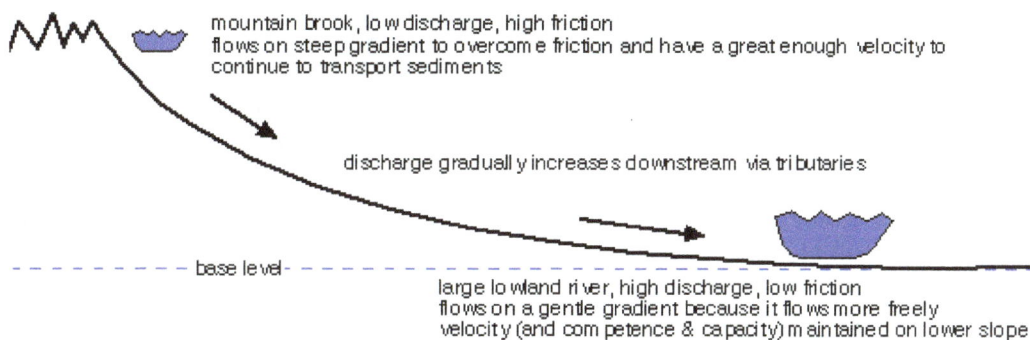

mountain brook, low discharge, high friction
flows on steep gradient to overcome friction and have a great enough velocity to continue to transport sediments

discharge gradually increases downstream via tributaries

base level

large lowland river, high discharge, low friction
flows on a gentle gradient because it flows more freely
velocity (and competence & capacity) maintained on lower slope

It seems counter-intuitive but stream velocity generally doesn't decrease on average, on the large scale from the steep headlands to the flat plains, from the dashing mountain brook to the broad peaceful river. The broad lowland rivers have much greater discharge and hydraulic radius. They flow much more freely (e.g., the water doesn't have to dash around boulders in the stream). The net result is that velocity actually increases somewhat.

Braided Stream patterns are found where there is a very large bed load where there is either a high sediment supply or the stream lies on a loose, unconsolidated bed of sand and gravel. In braided streams the stream does not occupy a single channel but the flow is diverted into many separate ribbons of water with sand bars between.

Stream Valley Evolution

Youthful Stream Valleys have steep-sloping, V-shaped valleys and little or no flat land next to the stream channel in the valley bottom.

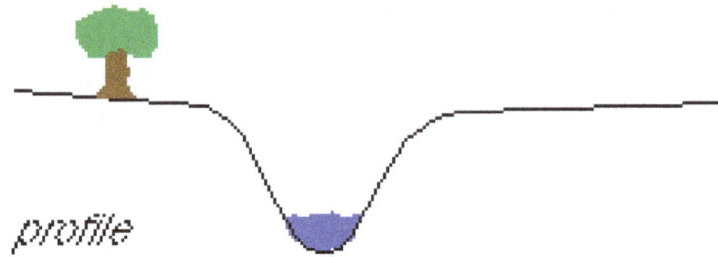

Mature Stream Valleys have gentle slopes and a flood plain; the meander belt width equals the flood plain width.

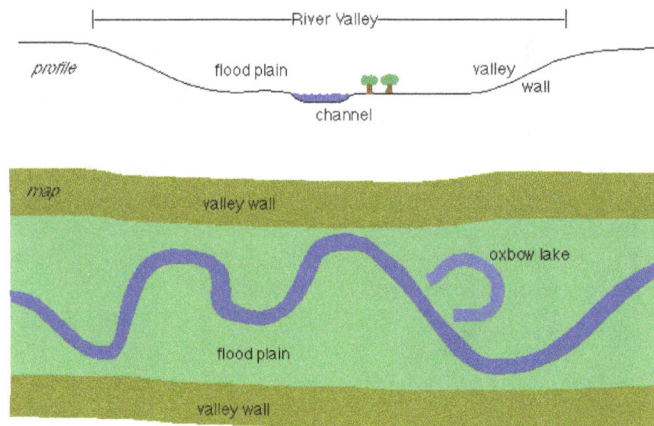

Old Age Stream Valleys have very subdued topography and very broad flood plains; the flood plain width is greater than the meander belt width.

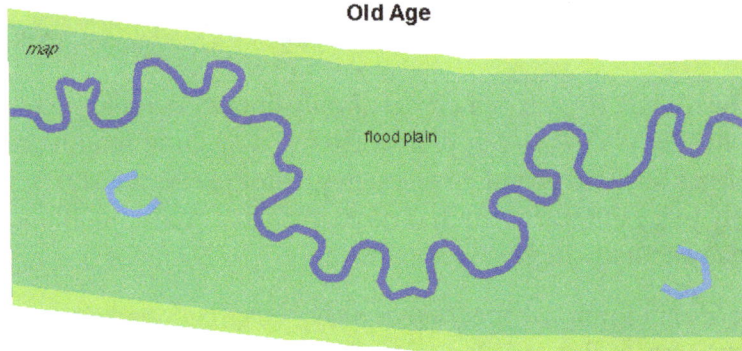

Stream Orders

Stream order is a measure of the relative size of streams. The smallest tributaries are referred to as first-order streams, while the largest river in the world, the Amazon, is a twelfth-order waterway. First- through third-order streams are called headwater streams. Over 80% of the total length of Earth's waterways are headwater streams. Streams classified as fourth- through sixth-order are considered medium streams. A stream that is seventh-order or larger constitutes a river.

When diagramming stream order, scientists begin by identifying the first-order streams in a watershed. First-order streams are perennial streams that carry water throughout the year and have no permanently flowing tributaries. This means no other streams "feed" them.

Once the first order streams are identified, scientists look for intersections between streams. When two first-order streams come together, they form a second-order stream. When two second-order streams come together, they form a third-order stream. And so on. However, if a first-order stream joins a second-order stream, the latter remains a second-order stream. It is not until one stream combines with another stream of the same order that the resulting stream increases by an order of magnitude.

Stream Ordering

Examining the stream network is important in determining study sites. It is best to sample a stream above and below any point at which a tributary enters it, as well as in the tributary itself. The result is 3 sample sites at each intersection of two streams. This is done so that one can narrow down the location of any potential pollutants.

Stream order is also an important part of the River Continuum Concept. The River Continuum Concept is a model used to determine the biotic community expected in a stream based on the size of the stream itself. As water travels from headwater streams toward the mouths of mighty rivers, the width, depth, and velocity of the waterways gradually increase. The amount of water they discharge also increases. These physical characteristics dictate the types of aquatic organisms that can inhabit a stream.

Classic Stream Order

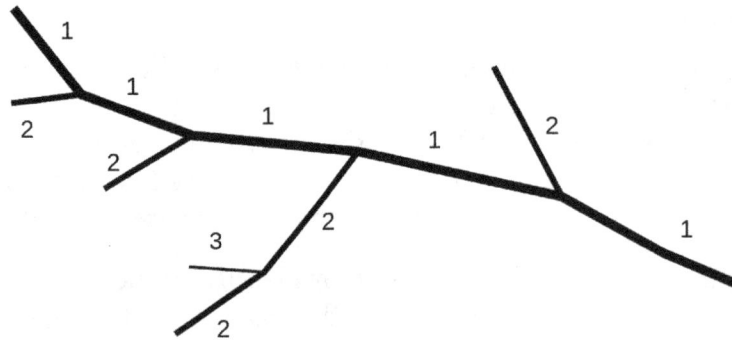

Classic stream order

The *classic stream order*, also called *Hack's stream order* or *Gravelius' stream order*, is a "bottom up" hierarchy that allocates the number "1" to the river with its mouth at the sea (the main stem). Its tributaries are given a number one greater than that of the river or stream into which they discharge. So, for example, all immediate tributaries of the main stem are given the number "2". Tributaries emptying into a "2" are given the number "3" and so on.

This stream order starting at the mouth indicates the river's place in the network. It is suitable for general cartographic purposes, but can pose problems because, at each confluence, a decision has to be made about which of the two branches is a continuation of the main channel, or whether the main channel has its source at the confluence of two other smaller streams. The first order stream is the one which, at each confluence, is the one with the greatest volumetric flow, which usually reflects the long-standing naming of rivers. Associated with this stream order system was the quest by geographers of the 19th century to find the "true" source of a river. In addition to the stream with the greatest length (the source at the maximum distance from the mouth) and the size of the various catchments, account was taken of the stream which deviated least at the actual confluence as well as the mere successive names of rivers such as the Rhine and the Aare or the Elbe and the Vltava.

Strahler Stream Order

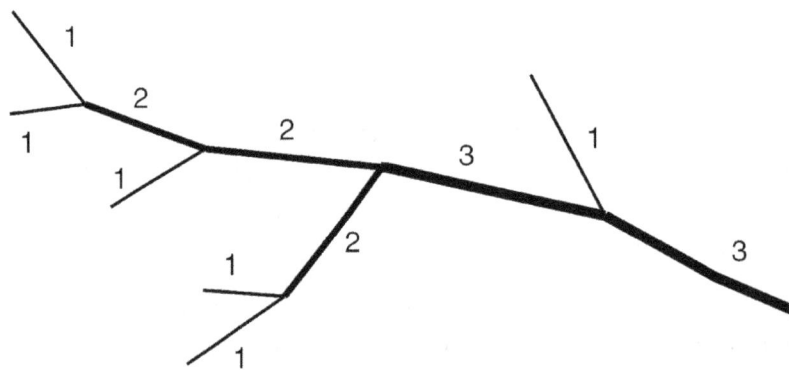

Strahler stream order

According to the "top down" system devised by Strahler, rivers of the first order are the outermost tributaries. If two streams of the same order merge, the resulting stream is given a number that is one higher. If two rivers with different stream orders merge, the resulting stream is given the higher of the two numbers.

Strahler order is designed for the morphology of a catchment and forms the basis of important hydrographical indicators of its structure, such as bifurcation ratio, drainage density and frequency. Its basis is the watershed line of the catchment. It is, however, scale-dependent. The larger the map scale, the more orders of stream may be revealed. A general lower boundary for the definition of a "stream" may be set by defining its width at the mouth or, by reference to the map, by limiting its extent. The system itself is also usable for small-scale structures.

Shreve Stream Order

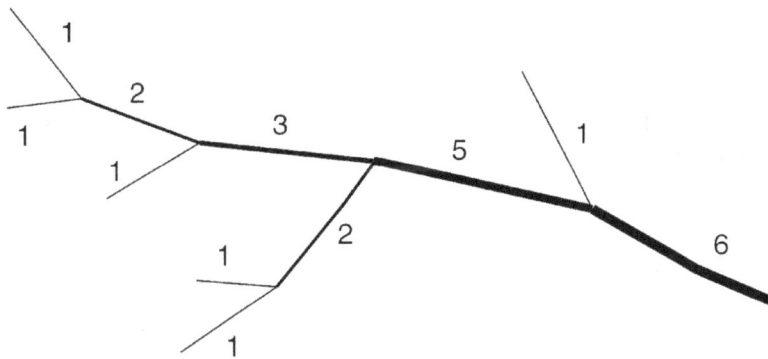

Figure: Shreve stream order

The Shreve system also gave the outermost tributaries the number "1". At a confluence the numbers were added together, however.

Shreve stream order is preferred in hydrodynamics: it sums the number of sources in each catchment above a stream gauge or outflow, and correlates roughly to the discharge volumes and pollution levels. Like the Strahler method, it is dependent on the precision of the sources included, but less dependent on map scale. It can be made relatively scale-independent by using suitable normalisation and is then largely independent of an exact knowledge of the upper and lower courses of an area.

Horton Stream Order and Other systems

Other systems include: the *Horton stream order*, an early, top down, system devised by Robert E. Horton; and the *topological stream order* system which is bottom up and where the stream order number increased by one at every confluence.

Classic Stream Order vs Horton & Strahler Methods

Classical or Topological ordering systems are assigned a dimensionless numerical order of "one" starting at the mouth of a stream which is its lowest elevation point. The vector order then increases as it traces upstream and converges with other smaller streams, resulting in a correlation of higher order numbers to the higher elevation of headwaters. Horton proposed to establish a reversal of this order. Horton's 1947 research report established a stream ordering method based on vector geometry. Five years later, in 1952, Arthur Strahler proposed a modification to Horton's method. Both Horton and Strahler methods established the assignment of the lowest order, number 1, starting at the river's headwater, which is the highest elevation point. Whereas Classical order number assignment correlates to height and elevation and traces upstream, Horton and Strahler's stream ordering methods correlate to gravity flow and trace downstream.

Both Horton's and Strahler's stream ordering methods rely on principles of vector point-line geometry. Horton's and Strahler's rules form the basis of programming algorithms which interpret map data as queried by Geographic Information Systems.

Usage

The classic use of stream order is in general hydrological cartography. Stream order systems are also important for the systematic mapping of a river system, enabling the clear labelling and ordering of streams.

The Strahler and Shreve methods are particularly valuable for the modelling and morphometric analysis of river systems, because they define each section of a river. That allows the network to be separated at each gauge or outflow into upstream and downstream regimes, and to classify these points. They are also used as a basis for modelling the water budget using storage models or time-related, precipitation-outflow models and the like.

In the GIS-based earth sciences these two models are used because they show the graphical extent of a river object.

Research activity post Strahler's 1952 report has focused on solving some challenges when converting two-dimensional maps into three-dimensional vector models. One challenge has been to convert rasterized pixel images of streams into vector format. Another problem has been that map scaling adjustments when using GIS may alter the stream classification by a factor or one or two orders. Depending on the scale of the GIS map, some fine detail of the tree structure of a river system can be lost.

Research efforts by private industry, universities and federal government agencies such as the EPA and USGS have combined resources and aligned focus to study these and other challenges. The principal intent is to standardize software and programming rules so GIS data is consistently reliable at any map scale. To this end, both the EPA and USGS have spearheaded standardization efforts, culminating in the creation of The National Map. Both federal agencies, as well as leading private industry software companies have adopted Horton's and Strahler's stream order vector principles as the basis for coding logic rules built into the standardized National Map software.

Stream Characteristics

The ability of a stream or river to erode and transport sediment is affected by many factors. These factors, which are interconnected, include the velocity of the water, the stream's gradient, its discharge, and the shape of its channel.

Velocity

The velocity of water in a stream or river is the distance that water travels in a given amount of time. The velocity of the water in a river is related to the amount of energy that the water has. A fast-moving river can erode materials more quickly and can carry larger particles than a slow-moving river. Many factors affect a river's velocity, including the steepness of the slope, the amount of water traveling downstream, and the shape of the path through which the water travels.

Gradient the steepness of the slope of a stream or river is called its gradient. A river's gradient varies along its course. A river may plunge down steep hills or mountains near its source. There its gradient is very large. By the time a river approaches sea level, it may be traveling across a plain that slopes very gradually, so its gradient is very small.

Gradient

The steepness of the slope of a stream or river is called its gradient. A river's gradient varies along its course. A river may plunge down steep hills or mountains near its source. There its gradient is very large. By the time a river approaches sea level, it may be traveling across a plain that slopes very gradually, so its gradient is very small.

Discharge

The discharge of a stream or river is the amount, or volume, of water that passes a certain point in a given amount of time. Discharge is not constant over the length of a river. In many rivers, discharge increases downstream because tributaries continually add more water. In rivers that flow into deserts, discharge may decrease downstream.

Discharge is not constant year-round. During times of increased precipitation or at times when snow is melting, more water runs into rivers. The velocity of the water also increases. Rivers become wider and deeper and may even flood their banks.

Current

The presence of definite and continuous current is the main characteristic of a stream. The current velocity, which may vary from 0.5 to 2 meter per second or more, depends on the stream gradient. The greater the discharge (volume/time), the greater the current velocity (distance/time) and also the amount of suspended material that is transported.

The rate of flow, in turn, influences a number of physical and chemical factors such as the temperature and dissolved oxygen concentration, which act directly on the biota. The current is a major limiting factor in streams. It might be assumed that plankton would be absent from fast- flowing streams, since such organisms are largely at the mercy of the current.

Plankton in small streams, if present, originates in lakes, ponds, or backwaters connected with streams and is soon destroyed as it passes through rapids of streams. Only in slow moving parts of streams and in large rivers, plankton is able to multiply and become an integral part of the community.

Land-Water Interchange

Streams are usually relatively shallow and, therefore, have a large surface compared to their depth. Land-water interchange is relatively more extensive in streams, resulting in a more open ecosystem. This means that the streams are more intimately connected with the surrounding land than are most lakes and ponds. For this reason a significant portion of the stream's nutrients falls into them from their banks in the form of terrestrial leaves, grasses and other debris.

In fact many of the primary consumers living in a stream feed on detritus and allochthonous material (i.e., outside the system) rather than on green aquatic vegetation (autochthonous production). The producers of a stream can supply only a fraction of the energy required by its animal consumers.

Oxygen

The third important characteristic of a stream, which is also a difference between a stream and a lake, is the high amount of dissolved oxygen. The reasons for this are many: flowing water, the relative shallowness of the ecosystem, and the large surface area exposed to the atmosphere. For these reasons the waters of a shallow, fast-moving stream will have higher oxygen levels than deep, sluggish rivers. In a stream the photosynthetic production of oxygen is not as important as it may be in a lake or pond.

Because of turbulence, the stream water is usually well aerated and problem of oxygen depletion, as seen in the hypolimnion of a deep lake, is almost absent. But stream animals are very sensitive to decreases in the oxygen content of the water. If a stream becomes polluted with excessive load of oxygen-demanding organic matter, such as the domestic sewage or the industrial waste, the resulting oxygen-depletion may cause serious problems such as a massive fish kill.

Stream Classification

Classification can vary along the stream and classification of a particular channel reach can vary through time. The most well-known classifications are based on stream pattern.

Classification Based on Stream Pattern

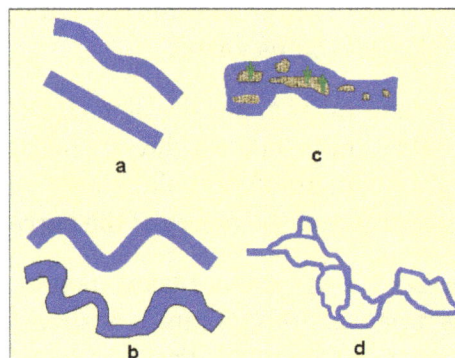

Figure: The different stream patterns: a) straight, b) meandering, c) braided and d) anastomosed.

Stream patterns describe the planform of a channel (planimetric view). The main characteristics used to separate patterns are the number of threads (single or multiple), sinuosity, and the island stability in a stream with multiple threads. Sinuosity is an important variable for describing stream meanders. It is the ratio of channel length to the straight length between the beginning and end of the same channel. Streams are considered meandering if sinuosity is higher than the ratio of 1.5.

The four main stream patterns are:

- Straight: streams with a single thread that is straight.

- Meandering: streams with a single thread but having many curves.

- Braided: streams have multiple threads with many sand bars that migrate frequently.

- Anastomosed: streams that have multiple threads but do not migrate laterally.

Classification Based on Stream Flow Conditions

Classification based on stream flow conditions are the result of the connectivity between stream and ground water. Perennial streams have water flowing in the channel year around and the stream is in direct contact with water table. Intermittent streams have water flowing only part of the year however, the stream is still in direct contact with water table. Interrupted streams have perennial water in their upper reaches and intermittent flow in their reaches at lower elevations. Finally, ephemeral streams flow with water only after precipitation events and the stream are well above the water table. Some large, permanent gullies may be considered ephemeral streams.

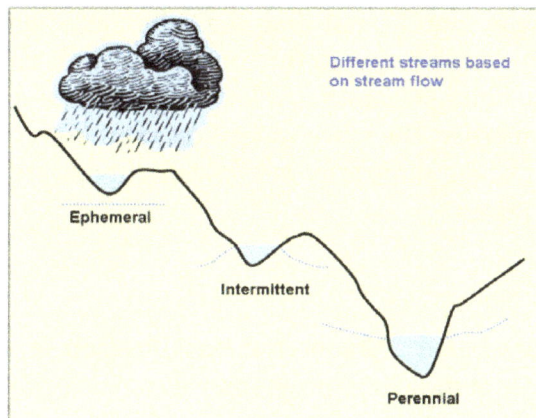

Figure: Ephemeral, intermittent and perennial streams.

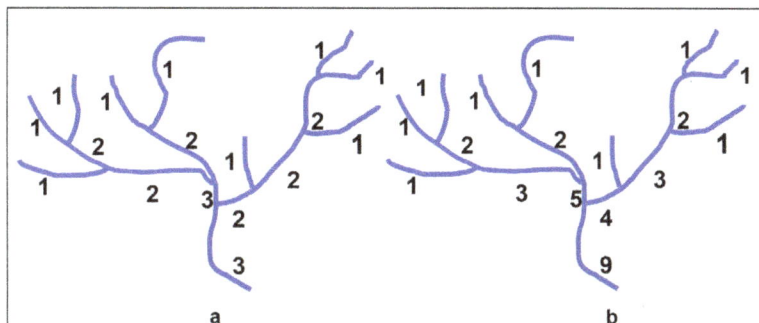

Figure: The two main stream order classification systems a.) Strahler and b.) Shreve

Classification Based on Stream Order

This classification is based on the type and number of tributaries that make up a channel network. Stream orders provide a way to rank and identify relative sizes of channels in a drainage basin. Smaller order numbers are given to smaller, headwater streams that are typically found in the upper reaches of a watershed. Higher order streams are given a larger number as the upper reaches join the main channel and are commonly found in the lowlands. The two major methods of ordering were developed by and named after researchers Strahler and Shreve. Both methods apply only to intermittent and perennial streams. The Strahler method is the most commonly used. First-order streams are the furthest upstream channels that have no tributaries.

A tributary is a stream that joins another stream reach or body of water. When two first-order streams unite, they form a second order stream. In the same way, when two second-order streams unite a third-order stream is created, and so on. Where two streams of different order join, for example a first and third-order, the combined stream retains the order of the higher order stream contributing to it. The main assumption behind this ordering system is that when two similar order streams join to create the next higher order stream, mean discharge capacity is doubled.

In the Shreve method, the stream order of the two streams contributing to a junction are added and provide the rank number of the stream below the junction. The rank of a particular stream represents the total number of first-order streams that have contributed to it. If we assume that the contributing area of each first order stream is approximately the same and that discharge is neither lost nor gained from any source other than the tributaries (which is not always true), then the Shreve number is roughly proportional to the magnitude of discharge in the stream to which it refers. For example, a first order stream has a lower magnitude discharge than a ninth order stream. The bigger the stream order, the more water flows through it. One of the shortcomings of stream orders is that streams of the same stream order might be very different if they are located in different ecosystems or climates.

Classification by Rosgen

Hydrologist Dave Rosgen developed a taxonomic classification of streams commonly used by land managers. It incorporates some of the previously described channel characteristics such as channel type, vegetation, and topography. His classification includes four levels. In this publication only Level I is described. The Level I classification integrates the following general geomorphic characteristics: basin relief, land-form and valley morphology. The specific characteristics are:

1. Entrenchment, which describes the relationship of a stream and its valley

2. Slope, which includes valley slope and sinuosity

3. Shape, which describes how narrow or deep, and how wide or shallow the channel is

4. Channel pattern.

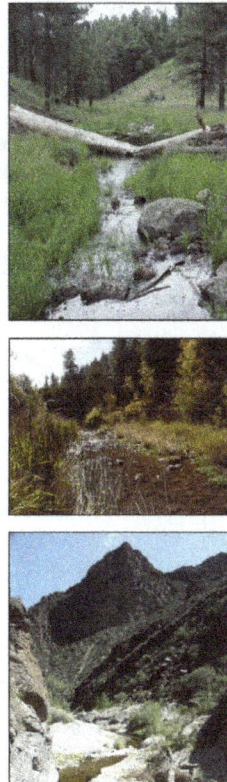

Figure: Examples of different Arizona stream types that can be classified by channel type, vegetation and topography (courtesy of R. Emanuel).

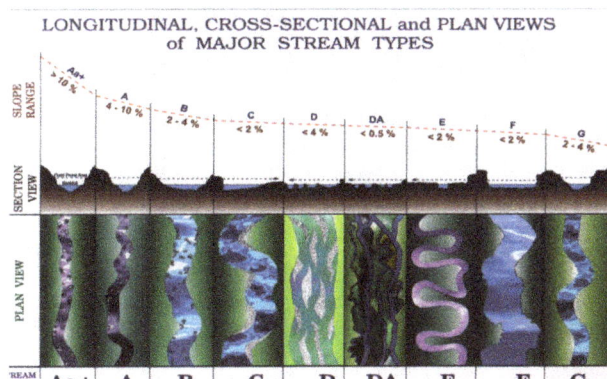

Figure: The nine types of streams based on Level I Rosgen classification.

Following Rosgen's Level I classification scheme there are nine stream types:

Aa+: Streams are very steep (slope > 10%), very entrenched, with low width/depth ratio (< 12) and totally confined laterally.

A: Similar to Aa+ but not as steep (slope 4-10%).

B: Streams are developed in moderately steep to gently sloped terrain (slope < 2%) with constricted valleys that do not allow a wide floodplain. Streams are entrenched, with moderate stream width/depth ratio (> 12), and low sinuosity (> 1.2).

C: Streams are constructed in alluvial deposits in narrow to wide valleys. These streams are sinuous (> 1.2), have low slopes (< 2%), slight entrenchment, and moderate width/depth ratio (>12) with a well-developed floodplain.

D: The main characteristic of a D type stream is the braided pattern with multiple threads.

DA: Streams of this type also have multiple threads, but in an anastomosed stream pattern.

E: Represents the end-point of channel stability for alluvial streams undergoing the channel evolution model sequence. Streams are slightly entrenched, highly sinuous (> 1.5) with a very low width/depth ratio (< 12).

F: Streams of this type are very deeply entrenched, in valleys of low relief with highly weathered or erodible material. Typically, these streams are working towards the re-establishment of a functional floodplain.

G: These streams are entrenched, narrow, and deep with low to moderate sinuosity (< 1.2). It is worthwhile to note that Rosgen's stream classification is not universally accepted. However, it does provide a framework for communication among stream managers, researchers, hydrologists, agency personnel and others interested in riparian issues. One of the primary criticisms of this classification method is that it does not incorporate the processes that created and will continue to modify the stream. In addition, the classification method has not been validated across all environments, and there is uncertainty of determining bankfull elevation and equilibrium conditions. Regardless of these shortcomings, the Rosgen classification is gaining popularity therefore stream and watershed stewards should be aware of this scheme.

Other Important Classifications

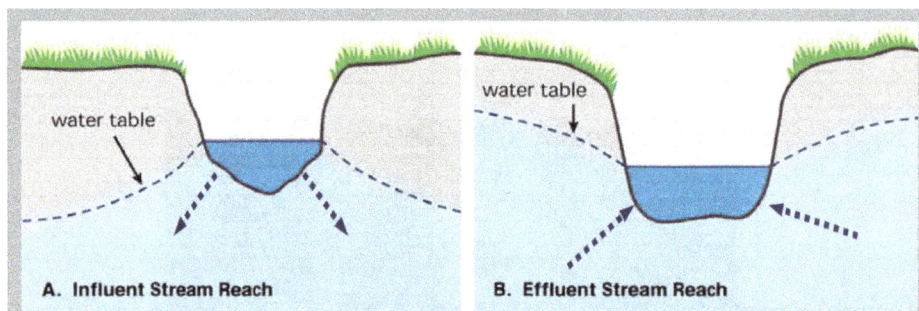

Figure: Gaining (effluent) and losing (influent streams) [from "Stream Corridor Restoration: Principles, Processes, and Practices, 10/98, by the Federal Interagency Stream Restoration Working Group (FISRWG)].

Regulated and non-regulated streams: Dams or other in-stream structures like dikes and levees control regulated streams. In contrast, non-regulated streams are not controlled by any human-made structures. Most streams in the United States are regulated. One of the last non-regulated streams is the upper reaches of the San Pedro River in southeast Arizona.

Channelized and non-channelized: Channelized streams or reaches have been artificially straightened while non-channelized reaches retain their natural sinuosity.

Channel material: Channel material may be classified as either alluvial or non-alluvial. Alluvial channels consist of material the stream has transported and can freely adjust its dimensions (such as size, shape and slope). Non-alluvial channels cannot adjust freely. An example of non-alluvial stream is one controlled by the geologic structure of bedrock.

Sediment load transport: The sediment load of a stream can be transported as suspended or bed load material. Suspended load consists of the sediments that are transported in suspension by turbulent flowing water. Bed load consists of the sediments that move along or near the streambed by sliding, rolling, or saltating. Saltating sediments are those that jump for a short time above the stream bed but return to the stream bed. Depending on which transport sediment method dominates, streams can be categorized as having suspended, bedload and mixed (where both suspended and bed load are equally important) sediment transport.

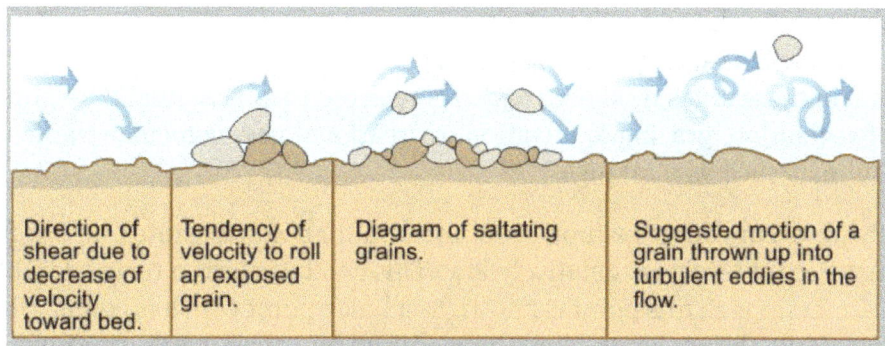

Figure: Sliding, rolling and saltating sediments on the stream bed

Stream Stability

Streams and rivers adjust their channel shapes and particle sizes in response to the supply of water and sediments from their drainage areas, and this in turn can affect streambed stability. Lower-than-expected streambed stability is associated with excess sedimentation, which may result from inputs of fine sediments from erosion—including erosion caused by human activities such as agriculture, road building, construction, and grazing. Unstable streambeds may also be caused by increases in flood magnitude or frequency resulting from hydrologic alterations. Lower-than-expected streambed stability may cause stressful ecological conditions when, for example, excessive amounts of fine, mobile sediments fill in the habitat spaces between stream cobbles and boulders. When coupled with increased stormflows, unstable streambeds may also lead to channel incision and arroyo formation, and can negatively affect benthic invertebrate communities and fish spawning. The opposite condition—an overly stable streambed—is less common, and generally reflects a lack of small sediment particles. Overly stable streambeds can result from reduced sediment supplies or stream flows or from prolonged conditions of high sediment transport without an increase in sediment supply.

Erosion Processes in Streams

The following processes that may occur singly, or in combination, are associated with stream erosion:

- overbank erosion

- bed erosion

- bank scouring (fluvial erosion)

- bank collapse/slumping (mass-failure)

- soil cracking and crumbling during dry periods

- channel avulsion (the development of a new or additional course for a stream)

Overbank Erosion

Bare ground on streambanks and adjacent riparian areas (often referred to as 'frontage' land in the grazing industry) will be susceptible to erosion by the action of raindrop impact. The complete removal of topsoil in these areas by sheet erosion is referred to as scalding. Bare ground can result from clearing of vegetation, grazing, cultivation, recreational use, vehicular traffic, weed control, fire and exotic animals such as feral pigs.

Rilling and gullying occurs where runoff from local rainfall concentrates before flowing over a streambank when stream levels are relatively low. This can be referred to as lateral bank erosion. The sharp increase in slope at this point is known as a knick point which is comparable to the head of a cascading waterfall. The concentration can be caused by a road, track, stock pad or an earthen bank or levee built to divert runoff from adjacent areas over a steep bank. The risk of lateral bank erosion is much greater when the stream is incised.

Dispersive clay soils are especially susceptible to erosion and are referred to as being sodic. They are common in many Queensland catchments and their floodplains. Sinkholes in dispersive soils can develop adjacent to a streambank and greatly accelerate the rate of erosion as they collapse and expand.

Sodic clays have sodium ions attached to the clay particles. When exposed to water, the size of the sodium ions increases and the links between individual clay platelets are broken causing the soil to disperse. Soils with a high proportion of silt can also act like dispersive soils. Particles resulting from dispersion remain suspended in water and are a primary cause of turbidity in dams, and the creeks and rivers flowing into coastal waters, western rivers and Lake Eyre.

Streambed Erosion

Bed erosion is the direct removal of material from the bed of the creek either by high velocity flows (causing uniform scour along the bed) or the formation of a head cut or 'knick point that migrates up the creek. Bed head cuts can be initiated as a scour hole in the streambed, which moves upstream as a result of the high velocities flowing over the cut. The upstream movement of the scour continues until equilibrium of the sediment and hydraulic regimes is achieved or until a scour resistant section of the bed is encountered.

Streambed erosion often has an innocuous appearance that gains minimal attention. But the lowering of the bed level is likely to lead to the destabilisation of streambanks, which is a much more obvious form of erosion which gains most attention.

Bed erosion in a low order 'stream' which is ephemeral and only carries runoff two or three times a year is really an example of gully erosion.

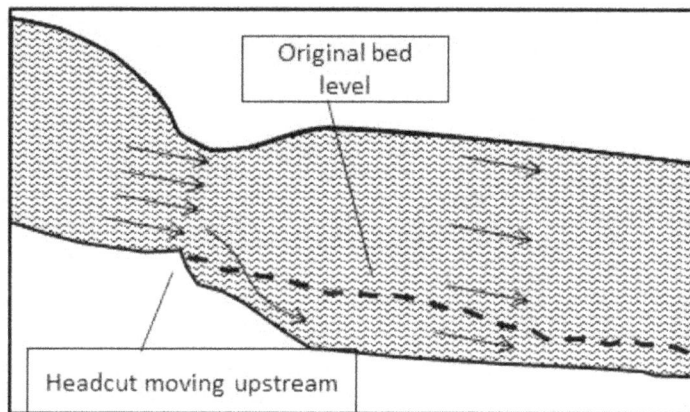

Figure: A head cut resulting from streambed erosion

Streambed erosion can be a natural process associated with down-cutting in a floodplain. The lowering of ocean levels during past geological eras was a significant factor contributing to the erosion of streambeds.

Processes and activities that can contribute to bed erosion include:

- A major flood following many years of sediment deposition in the streambed.

- An increase in bed slope caused by:

  o straightening the stream

  o removing a bed control such as a rock bar, weir or crossing or where reed beds are damaged by grazing stock

  o excavating the bed of a stream for extractive industries, recreation, large pump holes and dredging to deepen channels in streams and their estuaries.

- An increase in stream discharge as a result of:

  o increased runoff caused by catchment clearing and development

  o diversion of water from one catchment into another

  o prolonged flows delivering irrigation water

- An increase in velocity caused by:

  o a channel constriction caused by debris, fill, and vegetation on the riverbed

  o excessive desnagging and removal of vegetation from the channel

  o bridge abutments and culverts

  o levee construction

- A decrease in sediment supply by upstream dams, weirs, catchment erosion control measures, or excavations in the streambed can increase the risk of erosion. As the amount of bed load increases, the stability of the stream can be restored.

Figure: Excavating The Bed of a Stream Can Lead to Bed Erosion

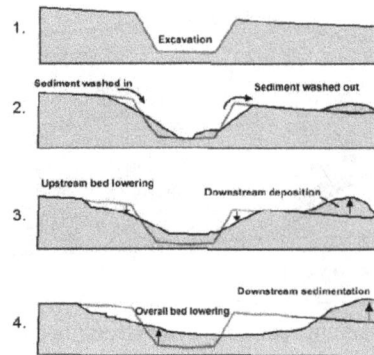

They included information on how extraction can sometimes be used strategically to rehabilitate streams affected by sediment slugs.

Bed lowering can have the following impacts:

- New gullies can develop or existing gullies can deepen in the overland flow paths leading towards the stream.

- banks are more susceptible to collapse because the height of the banks relative to the bed has been increased

- lowering of the groundwater level in the adjacent floodplain can adversely affect the aquifer and deny water to bores

- chains- of- ponds in streams can be drained

- cause a salinity problem by exposing saline groundwater aquifers

- access to water for pumps for irrigation and/or domestic supplies can be restricted

- cause downstream siltation which can destroy aquatic habitats and have adverse impacts on water quality, water availability, flooding, navigation and recreational pursuits

- result in damage to infrastructure including bridges, crossings and pumps

- reduced habitat for in-stream fauna such as fish and platypus.

The following features may indicate that bed erosion is occurring:

- vertical headcuts ('knickpoints') in the streambed

- steep and mobile riffles

- extensive bank erosion on both sides of the river

- loss of pools within the stream

- past attempts to stabilise streambanks may be well above normal stream level

- marks on bridge pylons that indicate the old bed level

- lateral erosion indicated by headcuts on tributaries (hanging valleys) and gullies

- a change in channel width between disturbed and undisturbed reaches

- exposure of 'ancient' logs and rock bars in the streambed

- wider, shallower reaches downstream of a headcut and fewer deep holes

- the complete relocation of a creek or a river

- sedimentation in the channel downstream.

- Falling water tables in nearby wells and bores.

Bank Scour (Fluvial Erosion)

Bank scour or fluvial erosion is the direct removal of bank materials by the physical action of flowing water. The process can occur in any section of a stream as a result of flooding or persistent, low flows against saturated banks resulting in the undercutting of the bank toe as shown in figure below.

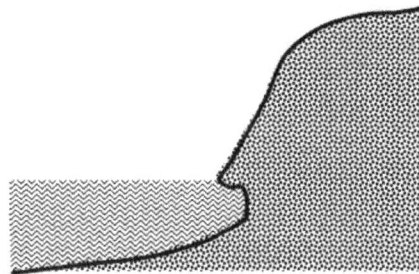

Figure: Undermining of a streambank by scouring of the toe-slope

The outside meander of banks can be especially susceptible to bank scour. Much of the sediment from this erosion can be deposited in a point bar on the next bend where the flow velocity is reduced. A helical flow pattern is established in the current by the alternate bars.

Figure: Helical flow currents acting on the outside of a meander bend

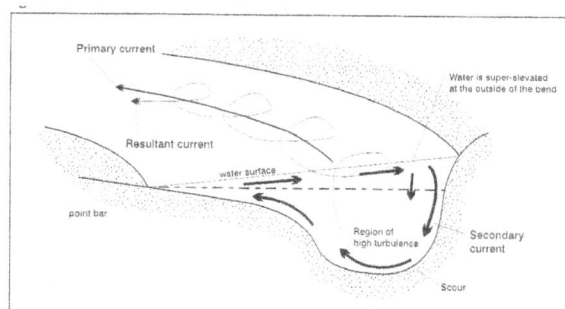

Bank scouring can occur as a result of the diversion of flows towards a bank by debris or vegetation in the streambed. The amount of roughness in a stream should be greater on the banks than on the bed of the stream and an excessive amount of vegetation growing in a streambed can divert flows against a streambank causing it to erode, the presence of woody plants growing in the middle of a streambed does not necessarily indicate that they are beneficial for the long-term functioning of the stream.

Wave erosion caused by wind or boat traffic can also cause scouring of the bank through a process known as fretting.

Streambanks are most susceptible to erosion when they are saturated. In wet soils, the amount of lubrication and pore water pressure between soil particles increases and soils become more erodible. Sandy soils are loosely bound and have very little resistance to flowing water. The high erodibility of unconsolidated sand is illustrated by the rapid loss of sand from beaches when exposed to high seas. Clay soils take longer to wet up and are more cohesive than sandy soils provided they are not dispersive. In clay soils, bank scouring could also be assisted by the process of slaking where soil aggregates break down in water due to the expulsion of trapped air from water sucked into pore spaces by capillary action.

Bank Collapse/Slumping

Bank collapse or slumping is where large chunks of bank material become unstable and topple into a stream under the influence of gravity. It is also referred to as mass failure, rotational failure or toppling failure and is comparable to a landslide. Bank slumping is often dominant in the lower reaches of large streams where there is no riparian vegetation and it can be the major source of sediment in flood flows. As slumping occurs, the streambank may reach stability by gradually battering itself back into the adjacent riparian land. Stabilisation measures such as bank reshaping, rock armouring a section of the bank and revegetation will reduce the distance over which this occurs.

Streambanks made up of sand and silt deposits are most susceptible to slumping because of their limited cohesion and sheer strength. If the bed of a stream is naturally armoured with rock which prevents bed cutting and deepening of the stream, serious bank erosion can occur during floods as this is the only opportunity for the stream to expand.

A precursor to bank slumping can be the formation of tensile cracks on a streambank during a dry period. Typically these cracks might be 10 to 20 cm wide at the surface and 1 to 2 m in depth.

Figure: Tension cracks on a streambank.

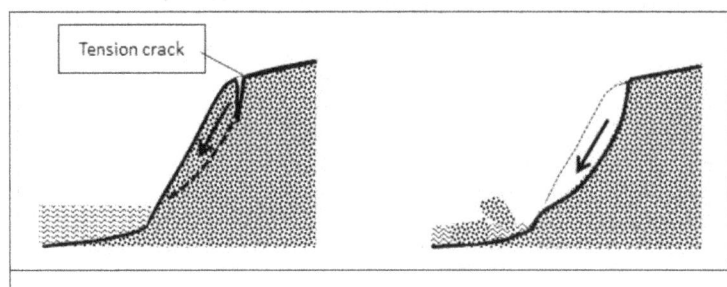

Another precursor to bank slumping can be erosion of the toe of a streambank by scouring

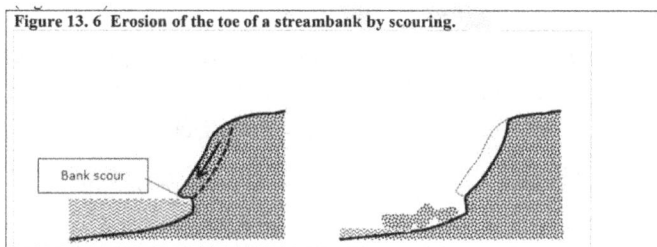

Figure: Erosion of the toe of a streambank by scouring

High water levels during a flood can infiltrate into dry streambanks for a distance of up to 2 m by capillary action. While the flood level remains high, there is some support for the bank but when the flood level recedes, slumping can occur because of the large amount of weight in the stream-bank. Saturated soils have low shear strength and slumping often occurs on the interface of the wet and dry zone. If the stream level in a flooded stream was to drop quickly as a result of closure of floodgates in a dam wall, slumping would be more likely to occur.

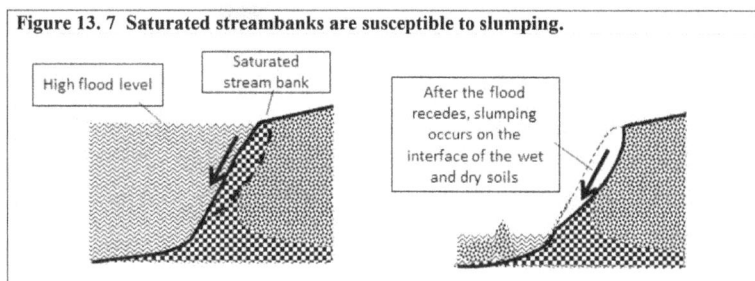

Figure: Saturated streambanks are susceptible to slumping

Seepage flows can also contribute to bank slumping and was considered to be a significant factor in the 2011 Lockyer valley floods. It was estimated that up to 10 000 cubic metres of soil was lost from several sites (Peter Pearce personal communication). A vein of very permeable sandy alluvium at depth in a streambank could allow seepage to exit at this point and lead to bank collapse. Possible causes of these seepage flows include former deep sand deposits from past flooding, septic tanks set to close to the embankment and overflow from domestic rainwater tank outlets.

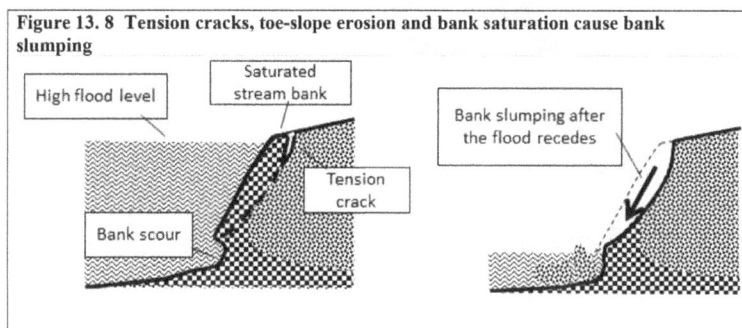

Figure: Tension cracks, toe-slope erosion and bank saturation cause bank slumping

The presence of tensile cracks, erosion of the toe-slope and saturation of the streambank can all combine to cause bank slumping as shown in figure below. Slumping often occurs a day or two after a flood has receded. Clear evidence of this is shown from the freshly exposed faces on the streambank that were not subject to flood flows. The slumped soil can remain at the base of the stream until it is removed as sediment during the next flood. Vegetation that was growing in the soil when it was on the bank can re-establish in the streambed.

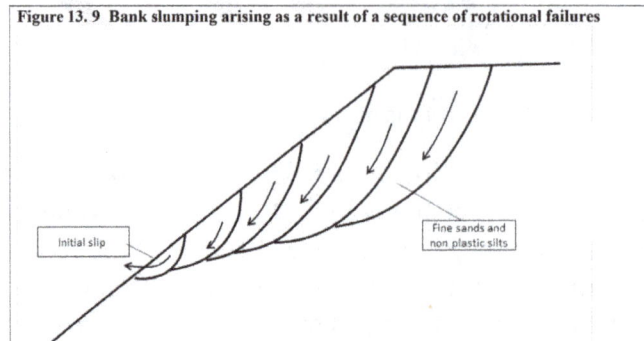

Figure: Bank slumping arising as a result of a sequence of rotational slumping

Kapitzke, referred to retrogressive bank movement as a variation of shallow rotational failure that occurs in fine sands and low cohesion silts, often in conjunction with seepage from the bank. The retrogressive failure starts with a small rotational failure in the region of the seepage face. This oversteepens the bank, initiating another rotational failure, and progresses until a lower, stable slope is achieved, or until the soil above the most recent slip has sufficient strength/cohesion to inhibit further failure. This explains how some apparently deep-seated failures have developed in slopes that otherwise appear too flat to be unstable.

Sidewall slumping is also a common process in gully formation in dispersive soils. Figure below shows how a shallow flow can cause undermining and subsequent collapse of a gully wall. In this situation, gullies will have a distinct 'U' shape. Such gullies can have very small catchments and will have often retreated to the very top of their catchments. They will only ever carry small flows even in a major rainfall event. Dispersive subsoils are infertile and support very little plant growth. They are consequently vulnerable to raindrop impact which can also remove significant quantities of soils from the sides and floor of these gullies. Structureless soils and soils that slake are also vulnerable to sidewall slumping.

Figure: Gully widening by undercutting and slumping

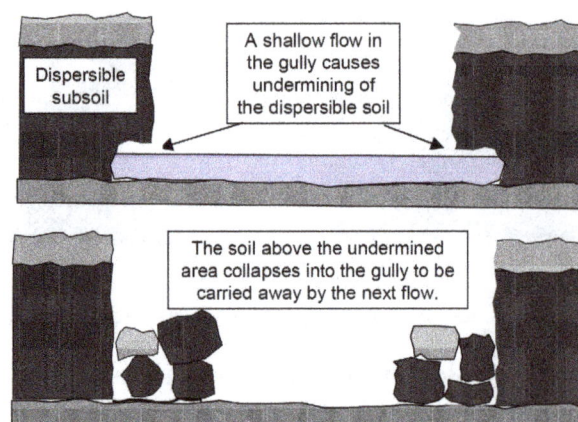

Channel Widening

Channel widening will occur as a result of channel incision and bank erosion. However, Brooks described other potential causes of channel widening:

- extreme floods; and

- a reduction in bank strength and/or roughness, caused by excessive stock grazing, vegetation clearing, or fire

Any suspected causal mechanisms should be addressed before considering any in-stream engineering approaches. Where widening is driven by rare high magnitude events, serious consideration should be given as to whether an interventionist engineering "solution" is really required. Assisted natural regeneration may be all that is required in these situations.

Soil Cracking and Crumbling

When soils dry out on exposed, clayey streambanks with a high shrink-swell potential, the formation of tension cracks may lead to slabs of soil falling into the streambed. As well as this process, a 3 to 4 cm layer of loose friable soil can fall away from the bank and form a scree at the base of the bank.

Channel Avulsion

Channel avulsion is the development of a new or additional stream course on a different part of the floodplain from the existing channel. it is natural for streams on a floodplain to relocate as they carry out their task of spreading sediment over the floodplain.

Channel avulsion is most likely to occur during a major flood. It can occur where there is some constriction in the channel flow, or following a period of significant sedimentation.

Any activity, or event, that lowers the height of streambanks or which redirects flood flows could lead to channel avulsion. Catchment activities that increase the rate and quantity of runoff and which increase sediment supply can also lead to avulsion.

Any concentration or diversion of flood flows on a floodplain caused by a road, levee bank or crop direction may lead to the development of a new channel. Dense weed infestation of stream channels will restrict flood flows and can contribute to deviation of floods out of the channel onto floodplains.

Stream Restoration

Stream restoration is a set of techniques or methods the County uses to protect adjacent properties and public infrastructure by reducing stream bank erosion, minimizing the down-cutting of stream bed, and restoring aquatic ecosystems (natural stream system).

Restoration techniques typically use natural materials such as rock, logs, and native plants to help slow down stormwater flow and restore the natural meander of curve pattern found in stable streams. They are usually done in larger scale projects utilizing large equipment to mobilize plants and rocks.

Stream Restoration Techniques

Brush Layering

Layers of live branch cuttings are placed horizontally along the stream. New plants will sprout from the live branches and the roots will hold the soil down and prevent erosion.

Coir Logs

Heavy mesh netting made from coconut fibers, used to hold soil in place and help plants grow, reduce weeds, and retain water. They naturally breakdown over time and become part of the soil.

Cross Vane

Stones are placed in streams in the shape of a "C" or a "V" to direct water towards the center of the stream away from the stream bank and reduce erosion.

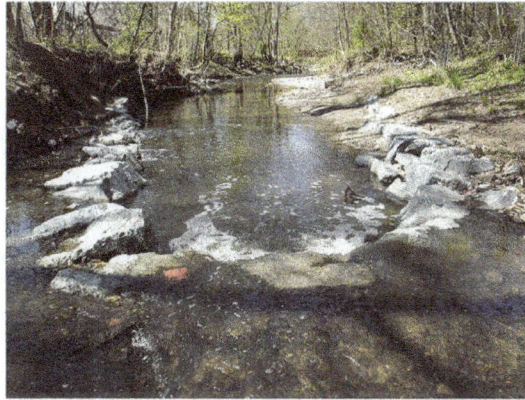

Grading and Planting

Steep stream banks are graded into a series of gently sloping steps. During large rain storms, the stream has more room for water to flow and decrease the speed of the flow. In addition, the plants and vegetation roots help stabilize the banks and hold them in place.

Imbricated Rip Rap

Stones are stacked together forming a wall or slope to prevent soil from being washed away into the stream during heavy rain storms. This is typically used in areas where erosion is severe or near private property.

J Hook

Rocks are placed in streams in the shape of a "J" to channel the flow of water away from eroding stream banks. The "hook" or curved tip of the "J" has slots for fast-flowing water to pass through and creates small pools (scour pools) where aquatic creatures can live.

Log Vane

Logs are placed and anchored to direct stream flow away from eroding stream banks towards the center of the stream. The concentrated stream current forms small pools (scour pools) below the vane where aquatic creatures can live.

Mulch Planting

Lengths of stream bank are planted with plants to stabilize the banks and hold soil in place through the roots of the plant.

Rock Pack and Flush Cut

Trees along the stream bank that are damaged by runoff can be protected with supportive rock packing. If the tree is beyond recovery, it can be cut down (Flush Cut) leaving the trunk to help hold soil in place.

Root Wads

Tree stumps with attached roots are anchored in stream banks with roots facing the streams to slow down flow and provide habitat for fish, amphibians, and aquatic insects.

Shallow Wetlands

A marshland-like environment is created below a storm drain outfall to allow treatment for the stormwater before it reaches the stream. It also provides for aquatic plant and animals.

Step Pools

A series of pools built with rocks that mimic staircase steps to slow down stream flow. This is often used to protect utilities such as sewer crossings, etc.

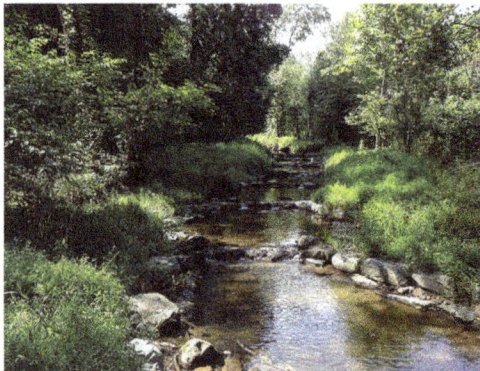

Stone Toe Protection

Large stones are placed at the base of the stream bank to prevent fast moving stormwater runoff from wearing away and destroying the stream bank. The stream banks can also be carved back to a gentler slope where native plants are planted to hold the soil in place.

Woody Debris

Woody debris includes logs and woody material, which can be used to provide spaces where fish can live and reproduce. Large tree limbs and woody materials are anchored along stream banks to reduce erosion and to buttress terraces and pools.

Permissions

Index

www.ingramcontent.com/pod-product-compliance
Lightning Source LLC
Chambersburg PA
CBHW082011190326
41458CB00010B/3151